Tourist Experience and Perception in the Pandemic and Post-Pandemic Era

Yuhua Luo · Xialei Duan · Xi Li
Editors

Tourist Experience and Perception in the Pandemic and Post-Pandemic Era

Springer

Editors
Yuhua Luo
University of the Balearic Islands
Palma, Spain

Xi Li
Faculty of International Tourism
and Management
City University of Macau
Macao, China

Xialei Duan
Faculty of International Tourism
and Management
City University of Macau
Macao, China

ISBN 978-981-96-1819-4 ISBN 978-981-96-1820-0 (eBook)
https://doi.org/10.1007/978-981-96-1820-0

This Springer imprint is published by the registered company Springer Nature Singapore Pte Ltd.
The registered company address is: 152 Beach Road, #21-01/04 Gateway East, Singapore 189721, Singapore

Preface

This proceedings book is a collection of the selected papers of the 11th International Conference on Tourism and Hospitality between China and Spain (www.china-spain.org) held in the City University of Macau both online and on-site, November 3–5, 2022.

The conference was held when the COVID-19 pandemic was still very active. However, the tourism industry and the hospitality industry did not stop its development. Instead, the industry and the researchers faced the challenges bravely and tried to fight back.

The researchers and scholars looked into the situation in different aspects of the industry, the target areas, target society groups and find the key factors that affect the experience and perception in tourism during and after the pandemic. The papers in this book provide a snapshot of the research outcomes under this specific period of time.

The accessibility to tourism destinations is one of the key factors the researchers identified. It is vital for normal people during the special circumstance such as in a pandemic and for disabled individuals in a normal situation. There is a paper specially explains a special form of tourism, the metaverse cultural tourism which uses the emerging media "the metaverse" for cultural tourism. A paper also addresses the importance of accessibility of the tourism destinations that can affect the satisfaction and revisit intention. A paper along this direction is trying to measure the connection of the tourists with the nature in a particular type of destination wetland parks.

An obvious trend in the papers in this book is trying to quantify and formalize the perception for some abstract concepts of the tourists. A paper tries to measure the perception of a special kind of tourism—the local knowledge tourism. It abstracts the dimension and develops the measurement scale for the perceived value. Similarly, another paper tries to analyze the spatiotemporal impact on tourist destination network attention due to the celebrity endorsement.

Other topics involved in the papers of this book include the mechanism of decision-making behavior, destination image perception, influencing factors in different types of tourism participation as tourist and as the hosts in the tourism destinations.

We hope that the publication of this book can contribute to the joining effort of human being in this special moment of the human history of a pandemic. It may serve as a "snap-shot" record of this special moment for our future generations in many years to come.

The series of our conferences (www.china-spain.org) is aiming at facilitating the exchange of research and development experience in the tourism and its related industries. We welcome scholars, industrial practitioners and government administrators to join us. We will continue our journey with all for a common goal—to have a better world.

We would like to thank all the authors for submitting their papers to this conference. All the members of the program committee, organization committee and the City University of Macau have contributed to the success of this conference. We are deeply grateful for their support. Without their support, this conference would not have been possible.

Seattle, USA — Yuhua Luo
Macao, China — Xialei Duan
Macao, China — Xi Li

Contents

Accessible Tourism: Metaverse Cultural Tourism Breaks the Boundaries

Yize Wu and Liang Liu

Abstract This study aims to present the understanding of the concept and application of accessible cultural tourism in the metaverse, focusing on the connection between metaverse, cultural tourism, and barrier-free tourism. It also discusses the advantages and significance of combining these three elements to form a solution to overcome obstacles for the people with disabilities (PWD). Furthermore, the paper delves into the service design for metaverse tourism by presenting a service blueprint. The blueprint underscores the integration of metaverse concepts with cultural tourism hotspots and ensures the barrier-free access for all travelers.

Keywords Metaverse · Cultural tourism · Accessible tourism · PWD tourism barriers · Service blueprint

1 Introduction

The World Health Organization (WHO) declared a COVID-19 pandemic on March 11, 2020, which has had a profound impact on the global tourism industry, including cultural tourism. [1–5]. The implementation of social distancing, limitations on mobility and mandatory quarantine, among other measures to contain the virus, has resulted in the interruption of travel and leisure activities on a global scale [6]. The emergence of a contactless culture has become a prominent feature of the contemporary society [7].

However, despite the physical interruption of tourism, the tourism communication never stopped. Throughout the pandemic, the industry continued to use virtual tourism technologies to influence the visitor's choices and promote its product [8]. As a pioneer of innovation, the metaverse has entered our vision in recent years. Its concept was proposed as early as the beginning of the last century. Along the years, the related classic movie *The Matrix* also deeply rooted in the hearts of many people. In recent years, high-quality works like *Ready Player One* have helped us to imagine

Y. Wu · L. Liu (✉)
Faculty of International Tourism and Management, City University of Macau, Taipa, Macau
e-mail: T23092100329@cityu.edu.mo

Y. Luo et al. (eds.), *Tourist Experience and Perception in the Pandemic and Post-Pandemic Era*, https://doi.org/10.1007/978-981-96-1820-0_1

the metaverse world again. We are no strangers to the world of the metaverse, and even look forward to it.

Industrial upgrading and tourism innovation are the development directions pursued by the tourism industry. The tourism industry needs to cultivate new product forms, production methods and consumption patterns. In the current coexistence with the epidemic, a travel method with many advantages such as metaverse tourism has also begun to attract everyone's attention. A famous tourist destination Zhangjiajie set up a metaverse research center at the end of 2021. It dedicates to the research and exploration of the integrated development of tourism and the metaverse, and drive the application and industrial innovation with the technological innovation. According to the survey [9] more than 84% of consumers are interested in using virtual reality (VR) or augmented reality (AR) for their travel experience, and 42% believe that the virtual reality and augmented reality are the future of travel. At present, smart tourism and VR tourism are booming, and it is expected to become a technological fashion. Using virtual technology to visit the great rivers and mountains can be said to complement each other. It is more meaningful and more realistic than what we imagined before.

Most of the researchers use the metaverse in tourism is for resources and collaborative creation. When the traditional stock resources are approaching exhaustion, the technical support of metaverse provides a new development direction for the cultural tourism industry. The application of metaverse technology in cultural tourism promotion places and the supply of its open-source creation and development have provided great support for the new development of the cultural tourism industry. The mode of new media dissemination, multi-person collaboration, and the integration of virtual and reality, as a creative platform for young groups, has truly realized a booming image of the new cultural tourism and boosted the development of the new cultural tourism.

In June 2022, the metaverse project jointly created by Datang Evernight City, a subsidiary of Qujiang Cultural Tourism, and Taiyi Group was launched. This is the world's first metaverse project with the historical and cultural background of the Tang Dynasty, called Datang Kaiyuan. In addition to the scene construction of the virtual city, the project will also use NFT and blockchain technology to build an ancient Chang'an city with millions of residents, so that the prosperous Tang Dynasty can be presented again in the metaverse. At the same time, the Datang Kaiyuan series of digital collections were sold out in seconds after the launch. This is the first digital collection of 3D architectural models in Xi'an, which also shows the enthusiasm of consumers.

The concept of metaverse is not yet mature. The related technology is still in the experimental and developmental stage, and is still far from industrialized production and application. With the current technology, it is only possible to provide diverse visual and auditory feedback, as well as providing very limited tactile feedback [10]. This is incomparable to the fun gained from experiential tourism that focuses on diverse experiences in reality. The application of metaverse tourism to cultural tourism, on the other hand, is a choice that builds on its strengths and avoids its weaknesses. Cultural tourism destinations are often limited by reasons such as heritage conservation, which prevents them from providing tourists with more open monuments and artifacts. However, using 3D modeling equipment and technology, it is only necessary to sample the digital information once, so that the fragile and exquisite cultural relics and monuments can be permanently retained in the virtual world. And the metaverse will provide unlimited access without regard to wear and tear, and can be based on real-life scenarios and historical accounts of destinations and local inhabitants, visitors at the time, thousands of years ago. The virtual community in the metaverse will allow visitors to visit without being alone, and the close connection with the online community will address the social needs of visitors. This will also differentiate the experience from the traditional solo online visit. Such a design provides a practical solution for how tourism research can follow the process of technological development to solve practical problems, and there is a great potential for the combination of metaverse tourism and cultural tourism.

In the Three-Year Action Plan for the Innovative Development of the Metaverse Industry (2023–2025), the Chinese government proposes that, in the short term, a number of typical applications of the metaverse will be created, and a number of new businesses, modes, and formats will emerge in the fields of life consumption and public services. Local requirements for the development of metaverse also point out that meta-universe, as a new technology application, should be used as an entrance to the digital industry to drive the development of the real economy and provide methods for solving social problems [9]. This study will propose practical new ideas for the development of metaverse tourism. And it is extended to be applied to the care of the disadvantaged groups, using the virtual nature and decentralization of the virtual world to provide a better perspective for the study of accessible tourism.

2 Literature Review

2.1 Metaverse

The metaverse is a virtual world that exists parallel to reality. By integrating multiple technologies, this world is closely related to reality. It provides an immersive experience based on augmented reality technology, creates a mirror of the real world based on digital twin technology, builds an economic system based on blockchain technology, and tightly integrates the virtual world and the real world into the economic system, social system, and identity system, allowing Each user produces content and edits the world [10]. The metaverse is still an evolving concept, and we are still adding new meanings to it with our understanding. People build virtual identity mapping with metaverse worlds in pursuit of new experiences [11].

Metaverse is the breakthrough point for the integrated application and development of major cutting-edge technologies. It represents the development direction of future technology and industry. Its basic concepts, concepts, and values, especially high-fidelity, open source, decentralization, and sustainability. The four characteristics of development point out the general direction of liberalization, personalization, virtualization, and experience of future industrial development, and are of great significance for leading the development direction of related industries [12].

2.2 Metaverse Tourism

The COVID-19 pandemic and the war in Ukraine have once again demonstrated the vulnerability of tourism to external and internal shocks. And increased awareness of climate change may also influence individuals' decision-making about their destination. Traditional tourism is in urgent need of change. Experiences like shopping, visiting galleries and museums, attending concerts, visiting heritage sites, etc. can all be offered in the metaverse to reduce carbon footprint and save time and money. As the level of related technology improves, the lost experience will be negligible. Metaverse tourism proposes new options for tourism consumption, and consumers' choices to reduce time and spending to reduce carbon emissions highlight the importance of developing tourism in the virtual world [13].

Metaverse tourism can upload tourists' personal portraits in the scene where tourists want to go, creating virtual characters to participate in the scene interaction in real time. Recently, Italy's top football league announced it would screen AC Milan vs Fiorentina in the "Arch-rivals" metaverse, allowing fans to interact in Serie A virtual rooms.

The technologies brought by Metaverse Tourism will also play an important role in enhancing marketing, customer relations, communication, helping customers in their decision-making process and improving customer experience. With proper service design and planning, Metaverse Tourism is expected to become a pioneer in tourism development and innovation.

2.3 Metaverse Cultural Tourism

Metaverse provides new possibilities for the reconstruction of virtual and reality at the level of cultural tourism. Differentiating the subject and identity of tourists is also the core of the cultural tourism experience, which also coincides with the concept of the metaverse. Every individual in the metaverse has a real identity that is different from the real world, and the isomerization of identity can be achieved by experiencing different lives in the metaverse, and tourists in cultural tourism are another person in the metaverse. An identity mapping. Therefore, it is the key to use emerging technologies such as VR and AR in cultural tourism to create a field where virtual reality is combined and scenes are blended. At the same time, in the scene of cross-cultural travel, virtual technology can be used to integrate and reshape various tourism resources, combine the virtual world with the real space, and give travelers a new experience.

At present, we can simply see the exotic scenery through live video, but this method not only has the defects of only two-dimensional plane display, but also has a big sensory difference from our personal travel. Using the technology of the metaverse to model the tourist attractions in 3D, imitating the three-dimensional scene, and at the same time with the sound effects, can make our travel experience more realistic, but this is not metaverse tourism. Metaverse tourism is not only a simulation of the form of tourism, but also includes creating a novel experience of tourism. For example, people's tourism can only be local tourism at that time, but in the metaverse, we can start a time journey, generate Past or future scenes, thus starting a journey across time. Here, we can return to the ancient battlefield and reach the future human world; of course, we can also go to the end of the vast world to explore the moon, walk Mars. This kind of tourism gives people a realistic experience in multi-sensory, and makes people feel truly immersive [14].

Among the many contents of cultural tourism, its cultural connotation and carrier are often difficult to express, display and copy. And the metaverse can just make up for this shortcoming. Metaverse tourism can break down the barriers of time and create scenes that travel through time and space. The items that need to be protected and treasured are copied through the corresponding modeling technology, and an unlimited number of tourists can enjoy it from multiple angles and without a time limit. It can also create a relevant atmosphere. And such a scene does not need to be carried out in a museum, in a historical monument. The atmosphere and feeling can be more in line with the times, which is undoubtedly a better choice.

2.4 *Metaverse Accessibility Cultural Tourism*

In current research, accessible tourism generally focuses on persons with disabilities. In a broad sense, accessibility tourism refers to within a certain area, based on maintaining the relatively complete autonomy of each region, by eliminating regional barriers, to achieve inter-regional tourism economic cooperation, so that tourism resources, products, markets, and information can be shared and realized. The elements of tourism flow are unimpeded. This study demonstrates the potential of meta-cosmic tourism to address accessible tourism in a narrow sense by addressing the problems of accessible groups. It also demonstrates the ability of meta-universe tourism to break down existing resource barriers, which is a big step towards reaching accessible tourism in the broader sense.

Accessibility design should be a common principle for service providers to design. But it is not well done. The general use of tourism customer journeys (TCJ) in research involves mapping a tourist's entire journey, outlining all the stages a person goes through when an experience is for tourism purposes [15]. The process that customers go through at all stages and touchpoints makes up the customer experience [16]. We applied the TCJ to the analysis of the situation of people with reduced mobility.

The four stages of TCJ include destination selection and tour design, journey to destination, stay at destination and transfer to another place or home. In addition to the above steps, to ensure the safety and care of PWD tourists in particular, these conditions must be guaranteed by a higher level of welfare system. Travel service providers must be able to work effectively with these regimes. But from our usual experience, we found that PWD is easy to be hurt or discriminated against during travel.

In the research on accessible tourism using TCJ theory, the barriers to tourism are summarized into the following six points: information barriers, architectural barriers, political barriers, cultural barriers, relational barriers, and technical barriers [17]. Information barriers refer to the difficulties people with disabilities face when choosing activities such as transportation, accommodation, itineraries, or shopping. Architectural obstacles refer to the actual obstacles encountered by PWD when using the means of transportation in the travel and using the sidewalk. Political barriers refer to the needs of PWD in the context of a political agenda. Cultural barriers: Refers to the indifference and neglect of the behavior of people with PWD by people of all abilities [18]. Relationship disorder refers to the fact that most people view PWD people from the perspective of self-experience, think that the PWD group is unique, and are unwilling to accept and help them. Technological barriers refer to the speed of technological change related to means of transport and communication, which can create a difficult situation for those who cannot keep up with the ongoing changes. The development of the epidemic problem in recent years is another round of new developments. Not having a smartphone, not using online programs, etc. are all social realities.

According to the researchers' description of metaverse tourism and accessibility tourism, both tourism for disabled people and accessibility tourism in a broad sense have a very good fit with the metaverse. Combining metaverse cultural tourism with general accessibility tourism can break down space barriers, allowing people with limited mobility to travel to tourist destinations without having to travel long distances. For accessibility tourism in a broad sense, this can also eliminate regional barriers and share tourism resources, products, markets, and information in combination with blockchain technology. For consumers, source information is open and transparent. It is also helpful for the healthy development of the tourism market and the formation of healthy competition. Combined with the above description, the metaverse breaks through multiple barriers and enters our neighborhood to be used, see Fig. 1.

In the face of the epidemic, although we lack the ability to completely solve the problem, our research is working hard for a better future. The application of metaverse accessibility cultural tourism can reduce or block related risks and contribute to the restoration of tourism vitality.

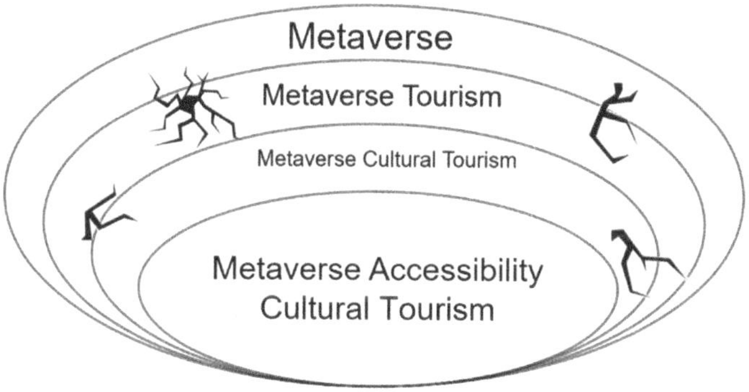

Fig. 1 Metaverse breaks multiple boundaries

3 Study of the Scenario of Accessibility Cultural Tourism in Metaverse

3.1 Study Approach

We propose the use of metaverse technologies in cultural tourism and can be well applied in accessible tourism (Metaverse Accessible Cultural Tourism). While it is currently in the initial stages of research like Metaverse Cultural Tourism, most current research articles utilize a speculative approach to predict the possible impact of Metaverse Cultural Tourism [11–15]. From the perspective of metaverse accessibility cultural tourism, there may be: How does accessibility tourism use the metaverse? What changes can the metaverse bring to accessible tourism? What real-world problems in accessible tourism can it solve, and how will it change the future of cultural tourism? We base our discourse analysis on the latest research on accessibility using a limited literature review. Starting from the six most significant problems of accessibility tourism, we will construct and analyze scenarios and solve relevant practical problems.

3.2 Metaverse Solves Obstacles

From the perspective of people with disabilities in tourism, we will use metaverse - related technologies and connotations to solve the six existing barriers in accessibility tourism.

3.2.1 Information Barriers

Lacking reliable information has been argued by a range of researchers to be a factor preventing the disabled from traveling [19, 20]. Individuals with disabilities encounter difficulties in assessing the risks associated with travel planning, travel to, and travel within a destination. These individuals require information about the specific assistance that is available to them, a finding that the study revealed to be often lacking on travel websites and travel agencies [21]. Given the information barriers in accessibility tourism, many situations in metaverse tourism directly avoid these problems. For example, when traveling in a virtual world, choosing a means of transportation is no longer difficult. PWDs don't need to worry about their physical flaws making it difficult to navigate the virtual world. The accommodation and difficulties encountered during the itinerary are relatively minor issues. In Metaverse Travel, users have the freedom to engage in any activity they choose. The metaverse is currently a highly popular phenomenon, with numerous countries and regions around the world proposing policies related to it [22]. Leveraging the popularity of the metaverse to promote accessibility tourism within the metaverse is undoubtedly the most popular approach for individuals with disabilities (PWDs).

3.2.2 Architectural Barriers

Metaverse tourism has obvious advantages in solving the problem of building obstacles. Hunt [23] has identified that tourism depends upon an inter-connected system of transport modes that is often not fully available to disabled people. Transport poses significant barriers to disabled travelers [24]. Airplane lavatories are often inaccessible to wheelchair users. This is a highly distressing and impassable barrier to disabled individuals traveling by air [25]. The general tourist classics lack accessibility facilities, and the design and construction costs of accessibility facilities are very high. But like the solution to information barriers, the metaverse is a virtual world where tourists use virtual bodies to tour, which will solve most of the facility barriers. And get rid of people who use metaverse devices at home. The construction barriers of PWD can be solved only by the construction of accessibility design by the institutions that provide metaverse tourism facilities. In this way, both the cost and the difficulty of implementation are greatly reduced. In fact, metaverse tourism will help the accessible community to ignore their physically bound state and gain liberation from their spirit and mind.

3.2.3 Political Barriers

Governments must adopt a culturally oriented approach to educate their residents to respect and accept the rights of people with disabilities. This should include integrating disability awareness and acceptance into broader social and cultural education programs [26]. It can be observed that a significant deficiency exists in the proactive implementation of web accessibility policies within the context of master plans for tourism services [27]. The protection of environments under national planning systems constrains the possibility of making necessary improvements to physical access. This indicates a conflict between conservation interests and accessibility requirements, suggesting that current legal frameworks do not fully accommodate the demands of accessible tourism. Changing the current political context is difficult, and when PWD tourists are immersed in the new metaverse travel world, both political theory and people's thinking are more advanced than they were in the past. Political obstacles are also easier to resolve.

3.2.4 Cultural Barriers

Despite the advances in technology, social attitudes and perceptions continue to represent a significant obstacle to the inclusion of PWDs in cultural tourism. Research indicates that tourism promotional materials frequently perpetuate a homogenous gaze that caters to non-disabled audiences, failing to adequately represent or accommodate the needs of PWDs [28]. Negative attitudes toward disabled people and a lack of cultural orientation toward respecting and accepting the rights of the disabled population are significant barriers [29]. The pre-trip phase is characterized by a lack of attention to the family unit and an intangible perception of the family tourism experience [30]. These cultural and attitudinal barriers contribute to a broader societal issue where people with disabilities are often marginalized, and their contributions and needs are overlooked. Metaverse travel frees the human mind from physical bondage, which will not be discussed in the meta-universe world, no matter what image you use. Even able-bodied people can use images of people with disabilities to experience this feeling of bondage and thus learn how to get along better with others.

3.2.5 Relationship Barriers

If you want to know some local information in an unfamiliar scene, you usually need to ask local people for help. In this process, our information source is completely based on the information provider, and the accuracy of this information data cannot be considered. The metaverse virtual environment gave the PWD population a valuable opportunity to re-socialize. In the currently used Metaverse Office, you can upload your villains to walk in a specific place to talk face-to-face or interact with each other. For those who don't want to communicate, you can mute measures, which is a very advanced and practical social function. It is hoped that there will be more technologies that can help our PWD people so that they can also communicate with others freely through their broken bodies. We can use the metaverse technology to set up NPCs in prominent places and ask NPCs for help to obtain the most real-time and effective data. Talking to metaverse NPCs with personalities can also help tourists with special needs [31]. In addition to human-type NPCs, conversations with animals and objects are also possible in the metaverse. Dialogue in the metaverse may be more exaggerated than because we can provide more expressions and emotions to please every metaverse visitor.

3.2.6 Technical Barriers

The metaverse represents an emerging digital space that offers a range of opportunities for accessible tourism. However, the lack of web design that is both universally accessible and adaptive to different user needs poses a significant barrier to PWDs [32]. Given the importance of web technologies in facilitating information access and service utilization for disabled citizens, this issue is particularly relevant [27]. Moreover, the absence of comprehensive guidelines or standards for accessible web design within the metaverse makes it more challenging to ensure equal access to cultural tourism options for all.

Technical barriers will be the most important part of the accessibility of metaverse tourism. The metaverse will be a collection of the most advanced technologies for many years. This can create a difficult situation for those who cannot keep up with the changes being made. The lack of smartphones, the inability to use online programs, etc. are all social realities. Of course, the difficulty of use does not always correlate with the difficulty of the technique. Building a simpler, more user-friendly human–computer interaction system is also part of the metaverse construction [33].

As a new idea to solve the problem of accessibility tourism, Metaverse Tourism has played a very good guiding role. We look forward to more research to enrich metaverse's accessibility to cultural tourism so that people in need have more and better choices when traveling.

4 Service Design

The service population of metaverse travel design is mainly the PWD population, the main target is the population with impaired mobility, and of course, it can be extended to the healthy population—especially people whose travel is restricted due to the epidemic. You can experience it by purchasing equipment or going to a designated service point. Table 1 below is a hardware comparison table the two metaverse service designs.

The construction of the metaverse service blueprint relies on technologies and research related to the metaverse. The tour process of tourists can be seen in Fig. 2, which is widely supported by metaverse-related technologies. And we can easily find that the number of service personnel needed is quite limited, and there are very few people in direct contact. This is also in line with the relatively safe travel method in the post-epidemic era, and it is also very friendly to the mental state of barrier-free groups.

Table 1 Comparison table of service hardware equipment

	Small family metaverse travel equipment	Metaverse travel service center
Introduction	Miniaturize the metaverse equipment and install it at home, and start the metaverse tour at any time by independent operation	With a full set of related equipment and professional assistance, experience the professional metaverse travel service
Easy to use	There are many types of equipment, and certain knowledge is required for operation and maintenance	Professionals operate and maintain, no relevant knowledge is required
Performance	For the sake of miniaturization and ease of use, the display effect is limited	Better performance can be achieved
Cost	The cost of equipment purchase is high, and the cost per use is low	Don't need to purchase equipment, and the cost per use is slightly higher
Convenience	Travel around the world without leaving home	Go to the Service Center for a full metaverse tour
Hygienic	No contact with other people is required	Access to a small number of people is required

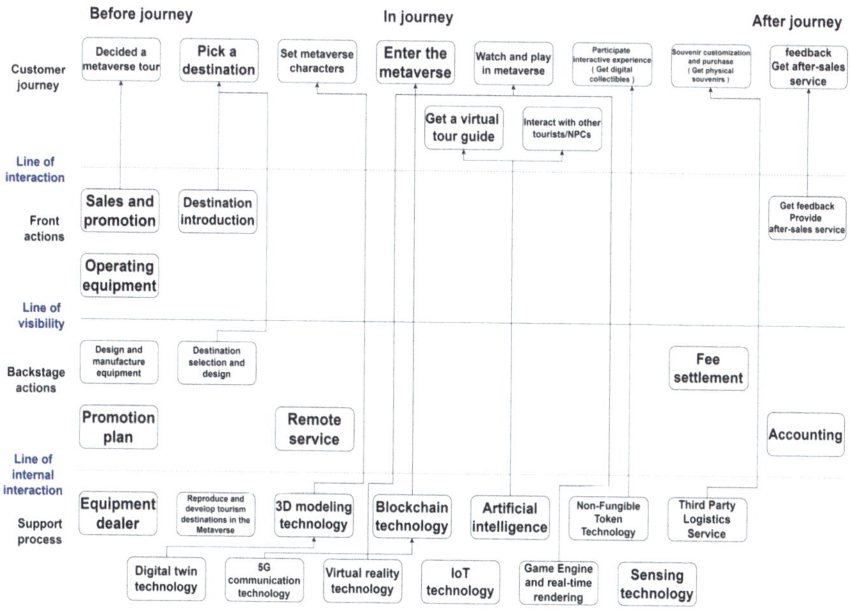

Fig. 2 Metaverse travel service blueprint

Specifically, the metaverse service blueprint shows us a framework for consumer and service provider behavior and service delivery before, during, and after the journey. Before the journey, both the consumer and the service provider need to do more work, but essentially all of it is out of sight and the level of contact is minimized. During the customer's enjoyment of the journey and service, reliable support from the service provider is important and relies on virtual contact. The customer value brought by the metaverse is not lost after the service, and the customer's journey will be continued through the online mall, for example. This is also an important part of the customer's subjective well-being, recall and satisfaction that cannot be ignored.

5 Summary and Criticism

Existing tourism barriers have greatly limited the development of barrier-free tourism activities. The six barriers studied in this paper are directly related to every PWD person. Metaverse concepts and new technologies the development of PWD is our direction to solve the obstacles, so this study introduces the metaverse technology to provide theoretical and technical support for the operation of barrier-free tourism for PWD people.

Although the proposal of new things will always experience various criticisms, from the perspective of the current social and technological development trends, the emergence of the metaverse is not far ahead. In the development process of the Internet, as the function center gradually changes from information to people, people's requirements for interaction are increasing, which drives the continuous progress of media technology and constantly changes the form of human society. The development of human society will also accelerate the continuous iteration of technology. The establishment of this development cycle allows us to predict the shape of the future society. The proposal of the metaverse is a kind of prediction and imagination of the future social form, and it is also an all-round reflection on the current social form. The idea of utilizing the metaverse and carrying out accessibility cultural tourism in this study is based on this idea.

With the increasing role of metaverse in improving the cultural tourism industry and the entire economic system, the risks contained in its technology integration, decentralization, high openness, and the integration of virtual and real are also more and more researchers and practitioners. In the application of the accessibility cultural tourism industry, does metaverse bring a sense of experience, content creation, an all-round improvement of interactive forms and PWD services, or does it lead to the wrong path of conceptual hype and virtual addiction?

Researcher systematically sorts out the challenges that the cultural tourism industry in the metaverse era may face from five aspects: monopoly risk, institutional risk, ethical risk, rights risk, and addiction risk [17]. Monopoly risk mainly means that under the control of large capital and large Internet companies, it is difficult for users to truly participate in the construction and content creation of the metaverse, and it is difficult to achieve complete openness and decentralization. Institutional risk mainly points out that due to the extremely complex monitoring system, the harm of disorder is extremely serious, and the regulatory agency is extremely imperfect. There is an essential contradiction between the decentralized open world concept of the metaverse of culture, tourism and the social organization structure and power operation mechanism in real life. The contradiction of ethical risk is more prominent. The ideal metaverse is a "Utopian-like" world that is highly free, highly open, and highly inclusive, but is highly isomorphic to the real world in terms of content generation, membership structure, and value orientation [18]. The rule system is an urgent problem to be solved. Rights risks are also particularly evident in the construction of the metaverse world. Blockchain technology can effectively solve the problems of information authenticity and security. But how to reasonably authorize and comply with the application in the service? How to achieve accountability? How to prevent new forms of crime based on data? These are new problems that follow new technologies. Addiction risk means that the high immersion and realism of the metaverse may make people addicted to the virtual world and ignore real life. Concerns about the metaverse have been around since day one of the concept of the metaverse. As for the accessibility metaverse cultural tourism, we synthesized the existing data and related research, combined with the actual situation, and believed that there was a public welfare risk in the project. The concept generation and specific development work of the metaverse are mainly controlled by big capital and large Internet companies. This makes it difficult to implement a project with certain public welfare attributes such as metaverse accessibility Cultural Tourism, and it is difficult to have an impact on the progress of technology and the direction of application. Only when the government helps from the level of laws and policies can it be developed and implemented easily.

Some people think that the metaverse is a new outlet for future technology and industrial revolution, some people regard the metaverse as a trap and a gimmick for capital expansion and hype, and some people regard the metaverse as a threat to human civilization drowning in comfort and losing its enterprising spirit. But anyway, here we propose the concept of accessibility cultural tourism in the metaverse. Undoubtedly, it is the emphasis on technology and human beings that will enable mankind to go farther. Research and product design serving tourism will reflect people's pursuit of a better life and a better future, which is also a direction with great potential for improving the quality of cultural tourism products and transforming and upgrading the tourism industry. Smart tourism, cultural tourism resource protection, social and public services and other fields will be given new vitality due to the close integration with the metaverse.

References

1. Junxiong, L., Nguyen, THH, Coca-Stefaniak, A.: Coronavirus impacts on post-pandemic planned travel behaviours. Ann. Tourism Res. 86: 102964 (2020). https://doi.org/10.1016/j.annals.2020.102964

2. Leach, M., MacGregor, H., Scoones, I., Wilkinson, A.: Post-pandemic transformations: how and why COVID-19 requires us to rethink development. World Dev. **138**, 105233 (2021). https://doi.org/10.1016/j.worlddev.2020.105233
3. Glusac, E.: "How Will Covid-19 Affect Future Travel Behavior? A Travel Crisis Expert Explains". The New York Times Company (2020). Retrieved from https://www.nytimes.com/2020/04/15/travel/q-and-a-coronavirus-travel.html
4. Kaczmarek, T., Perez, K., Demir, E., Zaremba, A.: How to survive a pandemic: the corporate resiliency of travel and leisure companies to the COVID-19 outbreak. Tour. Manage. **84**, 104281 (2021). https://doi.org/10.1016/j.tourman.2020.104281
5. Rasoolimanesh, S.M., Seyfi, S., Rastegar, R., Hall, C.M.: Destination image during the COVID-19 pandemic and future travel behavior: the moderating role of past experience. J. Destin. Mark. Manag. **21**, 100620 (2021). https://doi.org/10.1016/j.jdmm.2021.100620
6. Sigala, M.: Tourism and COVID-19: impacts and implications for advancing and resetting industry and research. J. Bus. Res. **117**, 312–321 (2020). https://doi.org/10.1016/j.jbusres.2020.06.015
7. Lee, H.K., Park, S., Lee, Y.: A proposal of virtual museum metaverse content for the MZ generation. Digital Creativity **33**(2), 1–17 (2022). https://doi.org/10.1080/14626268.2022.2063903
8. Lu, J., Xiao, X., Xu, Z., Wang, C., Zhang, M., Zhou, Y.: The potential of virtual tourism in the recovery of tourism industry during the COVID-19 pandemic. Curr. Issue Tour. **25**(3), 441–457 (2021). https://doi.org/10.1080/13683500.2021.1959526
9. Han, D.-I., Tom Dieck, M.C., Jung, T.: User experience mode for augmented reality applications in urban heritage tourism. J. Herit. Tour. **13**(1), 46–61 (2018). https://doi.org/10.1080/1743873X.2016.1251931
10. Ning, H., Wang, H., Lin, Y., Wang, W., Dhelim, S., Farha, F., Ding, J., Daneshmand, M.: A survey on the metaverse: the state-of-the-art, technologies, applications, and challenges. IEEE Internet Things J. (2023). https://doi.org/10.1109/JIOT.2023.3278329
11. Åstrøm, J.K.: Why theming? Identifying the purposes of theming in tourism. J. Qual. Assur. Hosp. Tour. **21**(3), 245–266 (2020). https://doi.org/10.1080/1528008X.2019.1658147
12. Lemon, K.N., Verhoef, P.C.: Understanding customer experience throughout the customer journey. J. Mark. **80**(6), 69–96 (2016). https://doi.org/10.1509/jm.15.0420
13. Gursoy, D., Malodia, S., Dhir, A.: The metaverse in the hospitality and tourism industry: an overview of current trends and future research directions. J. Hosp. Market. Manag. **31**(5), 527–534 (2022). https://doi.org/10.1080/19368623.2022.2072504
14. 张昭, 林续廖, & 陈昭彤: 把握趋势, 秉持理性, 多元对话——"元宇宙, 虚拟化社会与跨文化传播" 青年学术研讨会述评. *新聞文化建設*(01),3-. CNKI:SUN:WWJH.0.2022-01-001 (2022)
15. Kwanya, T., Stilwell, C. and Underwood, P.: *Library 3.0: intelligent libraries and apomediation.* Elsevier (2014)
16. Ma, Z.: Human action recognition in smart cultural tourism based on fusion techniques of virtual reality and SOM neural network. Comput. Intell. Neurosci. **2021**, 12 (2021). https://doi.org/10.1155/2021/3495203
17. 温晓君, 赵燕, 李旭东, 王丰, 王鹰, & 吕军: 元宇宙产业链生态白皮书.https://13115299.s21i.faiusr.com/61/1/ABUIABA9GAAg_vu3lAYopvyI1QM.pdf. [4月23, 2024 (2022)
18. 石培华, 王屹君,李中:元宇宙在文旅领域的应用前景, 主要场景, 风险挑战, 模式路径与对策措施研究. *《广西师范大学学报》*(哲学社会科学版), 58(4), 98–116 (2022)
19. Darcy, S., Daruwalla, P.S.: The trouble with travel: people with disabilities and tourism. Soc. Altern. **18**(1), 41–46 (1999)
20. Stumbo, N.J., Pegg, S.: Travelers and tourists with disabilities: a matter of priorities and loyalties. Tour. Rev. Int. **8**(3), 195–209 (2005). https://doi.org/10.3727/154427205774791537
21. Fraser, D.: *Perceptions of Risks and Barriers to Participation in Tourism for the Disabled* (Doctoral dissertation, University of Plymouth) (2017). https://doi.org/10.24382/558
22. Cassia, F., Castellani, P., Rossato, C., Baccarani, C.: Finding a way towards high-quality, accessible tourism: the role of digital ecosystems. TQM J. **33**(1), 205–221 (2020). https://doi.org/10.1108/TQM-03-2020-0062

23. Hunt, S.M.: Cross-cultural issues in the use of socio-medical indicators. Health Policy **6**(2), 149–158 (1986)
24. Shaw, G. and Coles, T: Disability, holiday making and the tourism industry in the UK: a preliminary survey. Tourism Manage. 25(3), 397–403 (20040). https://doi.org/10.1016/S0261-5177(03)00139-0
25. Poria, Y., Reichel, A., Brandt, Y.: The flight experiences of people with disabilities: an exploratory study. J. Travel Res. **49**(2), 216–227 (2010). https://doi.org/10.1177/0047287509336477
26. Kamyabi, M., Alipour, H.: An investigation of the challenges faced by the disabled population and the implications for accessible tourism: evidence from a mediterranean destination. Sustainability **14**(8), 4702 (2022). https://doi.org/10.3390/su14084702
27. Altinay, Z., Saner, T., Bahçelerli, N.M., Altinay, F.: The role of social media tools: accessible tourism for disabled citizens. J. Educ. Technol. Soc. **19**(1), 89–99 (2016)
28. Benjamin, S., Bottone, E., Lee, M.: Beyond accessibility: exploring the representation of people with disabilities in tourism promotional materials. J. Sustain. Tour. **29**(2–3), 295–313 (2021). https://doi.org/10.1080/09669582.2020.1755295
29. Cloquet, I., Palomino, M., Shaw, G., Stephen, G., Taylor, T.: Disability, social inclusion and the marketing of tourist attractions. J. Sustain. Tour. **26**(2), 221–237 (2018). https://doi.org/10.1080/09669582.2017.1339710
30. Cloquet, I., Palomino, M., Shaw, G., Stephen, G., Taylor, T.: Disability, social inclusion and the marketing of tourist attractions. J. Sustain. Tour. **26**(2), 221–237 (2018). https://doi.org/10.1080/09669582.2017.1339710IF
31. Gillovic, B., McIntosh, A., Darcy, S., Cockburn-Wootten, C.: Enabling the language of accessible tourism. J. Sustain. Tour. **26**(4), 615–630 (2018). https://doi.org/10.1080/09669582.2017.1377209
32. Hano, K.: Accessible tourism: concepts and issues. Tourism Plan. Develop. **8**(4), 467–468 (2011). https://doi.org/10.1080/21568316.2011.603886
33. Gilbert, T.F.: Human competence: engineering worthy performance. John Wiley & Sons (2013)

The Impact of Environment Accessibility on Tourist Satisfaction and Revisit Intention

Qianqian Liao and Xialei Duan

Abstract The accessibility of the environment can help to save travelers time, energy and render them a better leisure experience. A multitude of scholars have proved that accessible outdoor environment potentially encouraged elder people's mobility, especially for those with functional limitations. The aim of this study is to examine the effects of accessible environment on tourists' travel satisfaction and revisit intention. The China Greater Bay Area has been chosen for the study. A total of 415 valid questionnaires distributed online are used for data analysis. Results show that there is a significant relationship between security of accessible environment and tourists' perceived level of helplessness related to travel. In addition, the results of the study also confirm that tourists' perceived level of helplessness in the area have significant impact on their travel satisfaction and revisit intention. Discussions, limitations and suggestions for future research are also presented.

Keywords Accessible environment · Perceived helplessness · Tourist satisfaction · Revisit intention

1 Introduction

The rapid development of the social economy has resulted in a substantial increase in tourism participation, with seniors and disabled individuals increasingly venturing beyond their residences to appreciate the scenic allure of diverse destinations..

Compared with people without disabilities, people with disabilities are more likely to suffer from stigma and discrimination. They lack proper health care and rehabilitation services. Difficulty in using transportation, inaccessibility to buildings, and

Q. Liao
Faculty of Economics and Management, Guangdong Institute of Arts and Science, Zhanjiang, Guangdong, People's Republic of China

X. Duan (✉)
Faculty of International Tourism and Management, City University of Macau, Avenida Padre Tomás Pereira Taipa, Macau, People's Republic of China
e-mail: sallyduan@cityu.edu.mo

Y. Luo et al. (eds.), *Tourist Experience and Perception in the Pandemic and Post-Pandemic Era*, https://doi.org/10.1007/978-981-96-1820-0_2

inaccessibility of information and communication technology also have very serious impacts on their daily lives.

Accessible environment construction helps to alleviate the plight of the disabled. The accessible design aims to use contemporary technology to plan and transform the environment, to provide convenient, comfortable and safe space, and to create an equal, participatory and inclusive environment for the disabled. With the progress of accessible environment construction, the beneficiaries of accessible design are not limited to the disabled group. Making accessible benefits everyone has become a new connotation of accessible design in the context of the new era. The elderly will become its primary benefit group. As the age increases, the physical function of the elderly will decline. They will suffer from varying degrees of impaired mobility, hearing impairment and vision impairment. The accessible design just solves the additional needs of the elderly due to the decline in physical function.

Even though the disabled and the elderly are restricted in movement, with the development of society and the updating of concepts, more and more disabled and elderly people choose to travel. The Open-Door Organization (ODO) of the United States conducts a study on travel for the disabled. The study finds that a typical adult with a disability costs US$430, which mean the travel expenditure of the disabled population is as high as US$27 billion in 2 years [1]. Adults with disabilities spend approximately $13.6 billion on travel each year. In addition to the disabled, the elderly are also a group that cannot be ignored in the tourism industry. The elderly is an opportunity market for the tourism industry because senior people have more spare time than any other age group, and their spending on leisure exceeds that of young people [2].

These two great potential tourism markets will undoubtedly cause active competition for tourist destinations. The construction of an accessible environment in tourist destinations can solve the problems of the disabled and the elderly, such as difficulties in moving, information imbalance, and threats to personal safety.

Accessible environment can potentially affect the development of tourism. Wennberg et al. [3] have proved that accessible outdoor environment potentially encourages older peoples' mobility, especially for those with functional limitations. However, previous studies have focused more on the design of an accessible environment and have not linked it to tourists' revisit intention and satisfaction. Whether the accessible environment affects tourists' satisfaction and revisit intention is still a problem that needs to be explored. Therefore, this study will attempt to fill the research gap. This paper helps to achieve the awareness on the role of accessible environment in travel destination by investigating the relationships among accessible environment, tourists' intention and satisfaction. It's beneficial for tourism managers of the Greater Bay Area to explore new tourism markets and develop new marketing strategies to attract different types of travelers.

The aim of this paper is to analyze the influence of accessible environment on tourists' travel satisfaction and revisit intention in the area. The objectives of this paper are:

(1) To identify the extent to which the accessible environment affects the tourists' perceived level of helplessness.
(2) To explore whether there is a significant relationship among tourists' perceived level of helplessness, tourists' travel satisfaction and revisit intention.
(3) To examine whether there is a significant relationship between tourists' travel satisfaction and revisit intention.
(4) To investigate whether there are differences between person with impaired physical mobility and person without impaired physical mobility towards accessible environment.
(5) To recommend improvements to accessible environment operation and marketing for tourism related operator of the area.

2 Literature Reviews

2.1 Accessible Environment

In 2002, the U.N. Economic and Social Commission for Asia and the Pacific (ESCAP) adopted a resolution called the Lake Biwa Millennium Platform for Action, which called for efforts to build an inclusive, accessible and rights-based society for people with disabilities in Asia and the Pacific in the twenty-first century. Inclusive, accessible and rights-based summarized the ultimate goal of building an accessible environment.

Inclusion, accessibility and rights-based encapsulated the changing paradigm of attitudes towards people with disabilities. The prevailing view in academia was to divide the conceptual model of treatment of persons with disabilities into a medical mode and a social mode, and to add a rights-based mode to these two modes.

Based on the right-based mode and the social mode, the research on the accessible environment appeared in the academic circle. It was a collection of many research fields. Different scholars had various interpretations of the concept of accessible environment.

Wu et al. [4] mentioned that the accessible environment could be divided into two components. One was called the physical environment, such as the design of the following aspects: signs, facilities and equipment, buildings, roads, leisure and education areas. The other was intangible cultural environment, such as the acceptance of service personnel, the degree of respect, the quality of service, the attitude of local people, etc. Accessible environments provided services for all residents. It was a space or set of services where all people can move freely and safely, regardless of age, gender or physical condition [5]. And it ensured all people could use it independently, without hindrance and with as much dignity as possible. Jiao et al. [6] pointed out the urban accessible environment should include transportation, outdoor environment, indoor environment and information environment. The arrangement of various accessible facilities needed to be systematic and coordinated with the overall concept and shape of urban architecture.

Walls et al. [7] proposed a conceptual framework which provided a comprehensive understanding of customer experience in the tourism and hospitality perspective. The framework identified a number of features that influence tourism experience, including physical environment and human interaction environment. Physical environment referred to those physical facilities and surroundings, while human interaction environment referred to primarily interactions between tourists and service providers.

Accessibility, usability, and security were the three most commonly discussed dimensions in an accessible environment. These three different terms were often inadvertently confused and replaced by synonyms. Accessibility, usability and security concepts all had a foundation or relevance to the ergonomic field. However, they all had different emphases in their particular areas.

2.2 *Accessibility*

In the research on accessible environment, accessibility was a relatively well-researched field. It was specified and operationalized in a number of ways, resulting in a broad range of definitions. Table 1 summarized the example of several well-known descriptions.

The term accessibility originally appeared to address the travel needs of persons with disabilities. Accessibility could be described as the "ability to gain access to" and profited from specific structures or entities. The philosophy centered on allowing individuals with disabilities or special needs to join the workforce through the use of assistive technology [11].

Accessibility was a multi-dimensional concept that consisted of different dimensions. Conceptualization proposed by Geurs, van Wee [12] stated that accessibility encompassed a transportation dimension (e.g., mode of transportation), a land development dimension (e.g., the building construction), a time-base dimension (e.g., time spent on the trip), and a subjective dimension (psychological and physical needs, abilities).

Previous research of accessibility mainly focused on transportation. Accessibility of transportation was fundamental because it enabled everyone to move freely with

Table 1 Definition of accessibility

Authors	Definition of accessibility
Hansen, 1959	The potential of opportunities for interaction
Dalvi and Martin 1976	The ease at which any land-use operation can be accessed from a given location through a specific transportation method
Burns 1979	Individual's right to choose whether or not to partake in various activities

Source Ref. [8–10]

the help of the transportation system and to travel to their destination without obstacles [13]. In addition to the accessibility of transportation, the accessibility of urban planning was also getting more and more attention. Rumetshofer, Wöß [14] tried to combine accessibility with tourism. Their research suggested if a tourist destination was accessible, it must provide suitable infrastructure, accessibility information, accessible tourism websites and accessible tourism information systems.

In summary, accessibility helped people with restricted mobility to travel, while ordinary people could also benefit from it.

2.3 Usability

Accessibility was concerned with prerequisite for basic Use, while usability was concerned with optimized use. Similar to universal design and accessible design, usable design was centered on the creation of products that were highly efficient and easy to use.

The International Organization for Standardization (ISO) 9241–11(1998) described usability as "the extent to which a product can be used by specified users to achieve particular aim with effectiveness, performance, and satisfaction in a certain context of usage [15]. It looked at products in terms of their support for various functions, use scenarios, activities, or workflows, and it dealt with methods for assessing the reliable, effective, efficient, sufficient and satisfactory use of a product in a certain context. Petrie, Bevan [16] concluded that usability was not defined in absolute terms, but was relative to the user, the target, and the context of use applicable to a particular environment. Usability also needed consider flexibility, learnability, and memorability.

Usability was commonly considered a competitive advantage in contrast to accessibility. As technology became more and more of a commodity to be taken for granted, the demand for usability continue to increase [17]. As stochastic technology evolved, usability was not only reflected in physical facilities, but also in software requirements. The use of information and communication systems was accepted as a fundamental human right in the United Nations Convention on the Rights of People with Disabilities (CRPD). Disabilities could be sensory (visual, hearing, etc.) or cognitive and physiological, and they could be transient or permanent [18]. Especially for the people with disabilities. They had very special requirements for the usability of web-based tourism information systems. This was due to the heterogeneity of their disabilities (vision, language, hearing, body, etc.).

As mentioned above, an accessible environment included hard environment and soft environment. The hard environment contained infrastructure and accessible equipment. The soft environment referred to the channels through which people obtained accessible information. If the usability of the hard environment and the soft environment were improved, the sense of helplessness brought by the unfamiliar environment during the travel of tourists would be likely to decrease.

2.4 Security

When the two aspects of accessibility and usability were resolved, what needed to be considered was the security of the accessible environment. Security, on one hand, referred to the safety participants feel about the setting, such as whether the hotel entrance was close to the room they were staying. On the other hand, it represented the secure feeling derived from the participants' confidence in tourism services [19]. Its requirements applied to the intended group of users of the product. There was a trend to expand the user base to all users, regardless of ability or age in public areas, the community could realize the safety of tourists through surveillance cameras or police patrolling in the area. This showed that adding some additional equipment would enhance people's sense of security [20]. For the disabled, installing a telephone or emergency bell in the bathroom could effectively avoid emergencies.

When it came to the security of accessible environments, the first thing that came to our mind was the security of public transportation. The security of public transportation could not be ignored, as it had the potential to affect travel behavior at every stage of the journey, from preparation before the trip, perception during the trip, to evaluation after the trip [21].

Apart from traffic security, the personal security of tourists was also an issue they attached great importance to. Beecroft, Pangbourne [22] provided a conceptual framework to support the interpretation of personal security, which interpreted personal security as a two-stage dimension that included confidence, safety and security. As three key factors of personal security, confidence, safety and security were independent and mutually supportive. Security, described as freedom from external threats such as terrorism, abuse, robbery, or threatening behavior; Safety, defined as freedom from the effects of incidents, such as traffic accidents, stable interchange conditions, and the prevention of slips and falls; and confidence, described as freedom from mental or physical fatigue and tension, for example, a smooth ride quality, and from fear, for example, from not knowing how to travel, from injury, or a hazard [22]. Thus, it could be seen that the security problem was affected by many factors.

Special groups such as the elderly, women, children, and the disabled were in an objectively disadvantaged position physically and psychologically. When they were in an unfamiliar environment, they would evaluate their own safety by observing the surroundings. If they were in an environment without security personnel and any security measures, people would perceive unsafety. Insecurity could take many forms. To illustrate, people would start to feel physically ill; When a person realized that he was in an unsafe situation, he might become more and more worried as this awareness increased; Or a precautionary instinct that occurred unconsciously when a person evaluated potential hazards and unsafe areas around them. Effective security measured could largely eliminate the feeling of helplessness of tourists.

2.5 Seligman's Helplessness Theory

The psychological construction of learned helplessness was proposed by Miller, Seligman [23], which had been widely applied in the fields of aging and mental disorders.

Theory of learned helplessness stated that if an individual was discouraged from behaving freely, they would become more determined to behave as they want. However, if they were consistently discouraged from doing so, they would believe that the desired behavior was beyond their abilities. Consequently, even the environment improved, they would feel powerless, preventing them from behaving in the manner they had expected.

This theory had been applied to many fields of research. However, it rarely appeared in tourism literature. Lee et al. [24] studied the mediating role of learned helplessness in the relationship between travel constraints and travel intention of disabled Korean travelers. Wen et al. [25] explored the influence of tourism constraints and negotiation efforts perceived by Chinese elderly outbound tourists on the acquired helplessness of outbound tourism. Wen [26] investigated the willingness of parents and young children to travel together through the theory of learned helplessness.

The theory of learned helplessness was highly relevant to the study on tourism area related to people with disabilities. Because it assisted people to further understand the disabled group's travel intentions and travel behavior. It also explained why people with disabilities still did not engage in tourism even after tourism destinations had reduced the physical barriers associated. While the vulnerable groups were usually thought that had the same desire as normal people counterparts to engage in tourism events and activities, they would face more direct and indirect constraints when traveling. This was due to their social structure as well as their physical limitations [27]. Individuals with disabilities had fewer chances to engage in independent travel programs. Even if they had already engaged in tourism events, they might face unavoidable barriers that hinder their ability to enjoy the overall experience. As a result, some people might lose their intention to travel and therefore learn helplessness from a traumatic experience, while others might be more vigilant in subsequent interaction with the tourism industry but still followed their intention to travel.

Previous studies hadn't directly explored the relationship between accessible environment and tourists perceived learned helplessness. Lee et al. [24] explored the relationship between travel constraints and learned helplessness. Their findings indicated a significant relationship between travel constraints and learned helplessness. Subsequently, research findings from Wen et al. [25] stated that there's a positive relationship between tourists' perceived travel constraints and learned helplessness. Ying et al. [28] examined the relationship between tourism constraints and learned helplessness and their research findings revealed travel constraints would increase tourists perceived learned helplessness. The common point of all these studies were that there was a direct negative relationship between travel constraint and the perceived learned helplessness of special groups, such as senior tourists, tourists with pets and people with disabilities.

In an unfamiliar tourist environment, tourism constraints would increase the level of perceived learned helplessness. Accessible environment promoted the guarantee of the right to travel of all people, especially special groups, and aimed to remove constraints to and restrictions on participation. The emergence of accessible environment helped to eliminate the constraints of tourism and helped tourists get rid of the limitations of physical conditions, allowing tourists to carry out tourism activities independently and freely. Therefore, this study speculated the accessible environment could reduce the level of perceived learned helplessness of tourists.

2.6 *Travel Satisfaction and Revisit Intention*

In addition, the literature review of this study covered two broad areas of interest: research on travel satisfaction and revisit intention. Previous literature had shown evidence of a direct link between travel satisfaction and intention to revisit a destination.

Travelers evaluated their previous travel decisions based on their level of satisfaction or dissatisfaction with the destination and decided whether to revisit the place or not [29]. Customer satisfaction was the result of comparing service performance with customer expectation [30]. Extending it to the tourism industry, tourist satisfaction was the result of the comparison between the tourist experience in the destination and the expectation of the destination [31].

Hui et al. [32] conducted a survey on tourist satisfaction in four different tourist destinations, their results highlighted the importance of attractions to Asian tourists' travel satisfaction. Hui et al. [33] further explored the tourist satisfaction of Asian tourists based on diversity and safety, scenic spots and culture, service, price, people and convenience. The findings revealed all these five factors had positive effects on tourists' travel satisfaction regardless of nationality.

Repeated visitation could provide a steady stream of revenue that enabled businesses and destinations to invest in new market development. Gyte, Phelps [34] observed that travel suppliers could increase revenues and reduce costs by reducing their reliance on the more difficult task of attracting new visitors. Rittichainuwat et al. [30] conducted research to identify the impact of travel satisfaction on the likelihood of travelers to revisit Thailand. Researchers concluded that the higher satisfaction travelers had toward their trip, the more likely they would revisit a travel destination. They also revealed the importance of "quality, service, value of lodging and restaurant" on tourists' revisit intention.

Lau, McKercher [35] argued that previous experience of a destination had a strong impact on travelers' motivation to revisit a travel destination. Campo-Martínez et al. [36] conducted a study on revisit visitors, and the results indicated overall satisfaction had the greatest impact on tourists' decision of whether to revisit a place. An et al. [29] examined the relationship between travel satisfaction and travelers repurchase intention. Their findings suggested travel satisfaction significantly affected repurchase intention.

Many scholars had explored the relationship between travel satisfaction and revisit intention. However, rarely studies explored the relationship between these two areas under the framework of the learned helplessness theory. Lee et al. [24] first applied this theory to investigate the relationship between travel constraints and intention.

Based on these previous studies and learned helplessness theory, this study examined whether there was a significant relationship between tourists' travel satisfaction and revisit intention.

3 Methodology

3.1 Theoretical Framework

Based on the literature reviews, this study makes a more comprehensive summary of the accessible environment (Fig. 1).

Soft environment and hard environment of the accessible environment were analyzed from three aspects: accessibility, usability and security. In terms of security, this study focused on personal security. At the same time, this paper introduced the learned helplessness theory to study whether the level of perceived helplessness of tourists had an impact on the accessibility, usability, security of the accessible environment, and tourists' travel satisfaction and revisit intention.

The research theoretical model of this paper was shown in Fig. 2.

Based on the discussion presented in the literature and conceptual model, six main hypotheses were developed for testing. The hypotheses were proposed as follow:

H1: There was a significant relationship between accessibility of accessible environment and tourists' perceived level of helplessness related to travel.

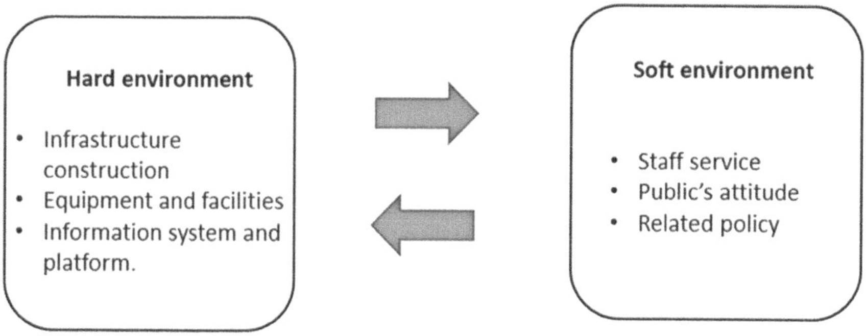

Fig. 1 Accessible environment

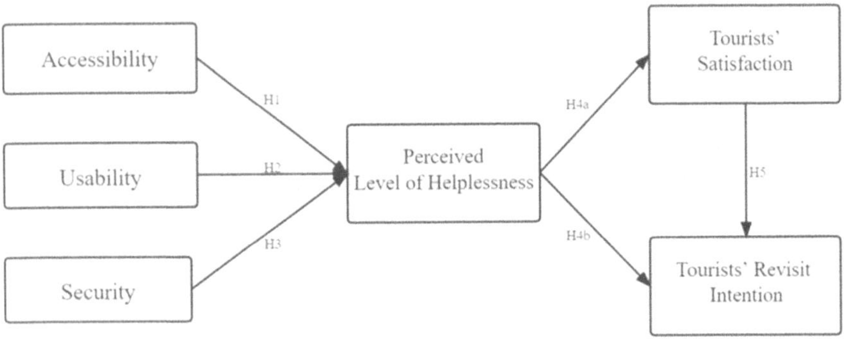

Fig. 2 Research conceptual model

H2: There was a significant relationship between usability of accessible environment and tourists' perceived level of helplessness related to travel.

H3: There was a significant relationship between security of accessible environment and tourists' perceived level of helplessness related to travel.

With the improvement of the accessibility, usability and security of the accessible environment, the level of helplessness perceived by tourists would decrease accordingly. Therefore, the following hypotheses were developed for testing:

H4a: There was a significant relationship between tourists' perceived level of helplessness and tourists' travel satisfaction with regard to Greater Bay Area.

H4b: There was a significant relationship between tourists' perceived level of helplessness tourists' revisit intention with regard to Greater Bay Area.

H5: There was a significant relationship between tourists' travel satisfaction and tourists' travel satisfaction with regard to Greater Bay Area.

3.2 Research Method

Research questions and topics usually dictated which method should be used to successfully complete the research. For the subject of this article, a quantitative method was selected. From the existing literature on this topic, it was not difficult to find that quantitative research methods were commonly used in tourism research related to the learned helplessness theory.

Lee et al. [24] used quantitative methods to study the perceived helplessness levels in tourism barriers and disabled travel intentions. Using quantitative methods in the form of surveys, a relatively large amount of data could be collected and converted into a database. A large number of statistical tests could then be performed on the information stored in the database to verify or reject these hypotheses. The advantage of quantitative research lies in the ability of researchers to make statements about the relationships between different participants by statistically testing the information collected from existing databases or surveys.

The quantitative research in this article would be able to provide information on the relationship between the accessible environment and tourists' travel satisfaction and revisit intention. In the future, if the government or related institutions want to conduct large-scale research related to it, they can use the relatively large amount of data collected from the survey in SPSS 25.0 to make general statements about the relevant population and not only the respondents.

In order to investigate these assumptions, it was necessary to identify and evaluate the opinions and attitudes of tourists during the research process. The research on accessible environments would inevitably involve the disabled and the elderly. Out of respect for them and taking care of their feelings, a suitable research method was needed to realize the quantitative nature of this research, so as to obtain the most true and effective data. Therefore, the main research tool was a structured closed three-part questionnaire.

The questionnaire was written in English and then translated into Chinese; method of two-way translation would be employed to evaluate the accuracy and reliability of the translation by tourism scholars.

The questionnaire was divided into three parts. The first part was the question of applicability. The interviewees of this study were all tourists who have been to the Greater Bay Area. The target research area of the study was Greater Bay Area. The second part allowed visitors to rate the accessibility, usability and security of the accessible environment in the target area. All 31 items were adapted to this research background and were based on a seven-point Likert- Like scale ranging from Strongly agree (1) to Strongly disagree (7). The third part was the demographic characteristics of the interviewee, such as gender, age, education level, and average household income.

Of the 31 items (Table 2), seven were used to measure accessibility, six were used to measure usability, four were used to measure security and five were used to measure levels of perceived helplessness. And there were four measures of travel satisfaction and five measures of revisit intention. The measurement items related to these three dimensions of perceived helplessness, travel satisfaction and revisit intention were based on previous research, and then modified according to the background of this paper [24, 37–40]. The measurement items of security were based on the theoretical framework of personal security [22]. Evaluating the security of the accessible environment from security, safety and confidence.

3.3 Data Collection

The target population of this thesis was tourists over 18 years old and had been to Greater Bay Area within two years. Even though previous studies based on the theory of learning helplessness all took the disabled, the elderly and other special groups as the research objects, this study attempted to extend the research objects to more groups.

Table 2 Measurement items of construct

Dimension	Items	Description
Accessibility	A1 A2 A3	1. Accessible facilities in GBA has nu effect on my travel experience, 2. The sidewalks in GBA have flat floors with no difference in height 3. The blind roads in GBA were paved and there are no obstacles to step them
	A4	4. There LS an accessible mode of transportation in GBA,
	A5	5. There are elevators, lifting platforms or accessible toilets in GBA 's public area
	A6	6. J can Innl accessible information about GBA on the Internet,
	A 7	7. Special personnel ire arranged for accessible guidance in GBA
Usability	U1	8. Everyone has access to the public accessible facilities in GBA
	U2	9. 1 don't need a lot of physical energy to use the accessible facilities in GBA
	U3	10. L have the flexibility to use the public accessible facilities in GBA
	U4	11. Accessible facilities in GBA are easy to use
	U5	12. There are Visual and textual instructions for Using accessible facilities in GBA
	U6	13. When I didn't know how to use accessible facilities, someone offered to help me,
Security	S1	14. It didn't break down when I used the accessible facilities m my travel to GBA,
	S2	15. I would feel comfortable when using accessible facilities in my travel to GBA
	S3	16. L did not experience violence or intimidation when using the accessible facilities in mv travel to GBA
	S4	17. The service staffs (hotel staff, shopping mall staflf, etc.) in GBA were well-trained, and they could help me solve an accident in
Level of percCiVcd helplessness (PH)	PH I	18. When [travel to GBA, 1 feel helpless
	PH 2	19. Before I came to GBA, I knew nothing about here
	PH 1	20. Traveling to GBA makes me in a gloomy mood
	PH 4	21. Comfortable traveling does not exist forme
	PH 5	22. Traveling to GBA is not a kind of thing for me to Cnjoy
Travel satisfaction (TS)	TS 1	23. 1 Was satisfied with my travel to GBA,
	TS 2	24. The accessible environment of GBA made my trip easier and more en lovable
	TS 3	25. My choice to travel GBA was a wise one
	TS 4	26. This experience is exactly what I expected
Revisit intention (RI)	R1 1	27. Whenever 1 have a chance to travel to GBA, 1 will
	R1 2	28. If [have a chance, [intend to revisit GBA in the next 12 months
	R1 3	29.1 predict 1 w ill revisit GBA in the future
	R1 4	30. In the following year, L may visit GBA again for tourism
	R1 5	31.1 wish to visit GBA again for tourism

According to Declaration on the Rights of Disabled Persons [41], disability was not a disease, but a condition in every person's life. Jia [42] also believed that everyone would have physical inconveniences in their life, including the disabled, the elderly, pregnant women and children, and even the able-bodied people holding heavy objects would have obstacles at that moment.

The aim of creating and improving an accessible environment was not only to remove the obstacles that prohibit people with disabilities from engaging in social life, but also to provided a social environment and living space for all members of society to freely travel and interact in different life stages.

Therefore, the study population was tourists aged 18 years and above, and they were subdivided into two groups: person with impaired physical mobility and people without impaired physical mobility.

Questionnaires were distributed to interviewees online. Convenience sampling was employed to select the respondents. The number of the questionnaires was expected to collect among three hundred and three hundred fifty, and the occupation of effective data should be more than 80%. Data would be gathered during the months of December in 2020 and January in 2021.

3.4 Data Analysis

Various statistical methods were used to investigate the relationships among the mentioned constructs. Firstly, The Statistical Package for Social Sciences (SPSS) 25.0 was used for descriptive analysis to provide respondents' profiles. Then, independent t test was employed to examine whether there were significant differences among the respondents with different demographic backgrounds in using accessible environment. Next, the partial least squares (PLS) analysis was employed to test the structure of theorical model via Smart PLS 3.3.3 software. We tested the theoretical model (validity and reliability of the measures) and then evaluated the structural model using the suggested SEM two-stage research technique [43]. The bootstrapping process (5000 resamples) was used to assess the relationship of path coefficient and load.

4 Results

Descriptive analysis (Table 3) was conducted to summarize the demographic information of the sample via SPSS 25.0.

The sample consisted of 415 respondents from all over China. 59% of respondents were with impaired physical mobility. The data indicated that the majority of them were aged between 36 to 45(28.4%), and most of the respondents had a bachelor degree(57.8%). The majority of the sample had an average household income of 5,001–7,000 RMB per month (30.8%).

Table 3 Demographic of respondents

Profile	Categories	Frequency	Percentage
Identity of interviewee	People with impaired physical mobility	245	59
	People without impaired physical	170	41
	Total	415	100
Gender	Male	175	42.2
	Female	240	57.8
	Total	415	100
Age	18–25	87	21
	26–35	88	21.2
	36–45	118	28.4
	46–55	74	17.8
	55–65	38	9.2
	66 or above	10	2.4
	Total	415	100
Education level	Less than secondary/high school	13	3.1
	Complete secondary/high school	79	19
	Bachelor degree	240	57.8
	Master degree	75	18.1
	Doctor degree	8	1.9
	Total	415	100
Average household monthly income	Less than 3,000	28	6.7
	3,001–5,000	83	20
	5,001–7,000	128	30.8
	7,001–10,000	102	24.6
	10,001 or above	74	17.8
	Total	415	100

Exploratory Factor Analysis (EFA) was conducted to reduce unnecessary items and increase the data's validity via SPSS 25.0. Satisfactory Kaiser–Meyer–Olkin (KMO) and Bartlett's Test of sphericity scores were used before proceeding to EFA. Hair et al. [44] suggested that the factor loadings of EFA should be 0.50 or greater. However, some items did not meet this standard or had a cross loading, and the variation between factor loading and cross loading was too small. So, these items were considered to be deleted. As shown in Table 4, U5 from usability had a bad factor loading. S1 and S2 from security had cross loadings with other variables, and the values of factor loading and cross loading were very close, which meant they also had strong correlations with other variables. In order to increase the accuracy of the items, these two would be deleted. Besides, A1 from accessibility could not be classified into any variables, it would also be deleted. Meanwhile, U6 from usability was found to have a strong correlation with accessibility, so it was classified into component 2 (Accessibility).

After deleting these four items, other factor loadings were all greater than 0.50, indicating the reliability of the research scale. The KMO value of each variable were greater than or equal to 0.5. The total KMO value was 0.94, indicating the items were marvelous [45]. The p-value of Bartlett's Test was 0.00, and the cumulative percentage of the total variance explained by the factors extracted was 80.56%. Other EFA results including component, mean, St. Deviation, KMO value and variance extracted of each item were shown in Table 5.

The measurement model was tested via SMART PLS 3.3.3 software. Firstly, factor loadings, composite reliability (CR), and average variance extracted (AVE) were used to assess the measurement model's convergent validity. Table 6 presented the factor loadings, CR values and AVE values.

According to Fornell, Larcker [46] criteria, the AVE values should be better than 0.50 (AVE > 0.50). As we can saw in Table 6, all the AVE values were greater than 0.5, ranging from 0.6 to 0.9. Therefore, the model converged with a satisfactory result.

The internal reliability of constructs was examined using composite reliability (CR) and Cronbach's Alpha. CR was best suited to PLS because it was sorted according to the reliability of variables, while CA was very sensitive to the number of variables in each construct [47].

For this study, Cronbach's alpha for each variable ranged from 0.719 to 0.97. CR values of each construct were between 0.88 to 0.98, which indicated strong composite reliability.

To determine the degree to which the construct did not correlate with other variables, discriminant validity was checked. Table 7 showed the square roots of the AVE of each construct. The AVE's square roots were greater than the correlations between the variables, indicating the adequate discriminant validity.

Some researchers had criticized the Fornell, Larcker [46] standards for failing to detect a lack of discriminant validity in academic research in a reliable manner. Some scholars had recommended an alternative approach to access discriminant validity:

Table 4 Deleted questionnaire items and reason

Variables	Items	Item description	Reason
Accessibility	A1	Accessible facilities in GBA have no effect on my travel experience	Do not belong to any variable
Usability	U5	There are visual and textual instructions for using accessible facilities in GBA	Bad factor loading value (less than 0.5)
Security	SI	It didn't break down when I used the accessible facilities in my travel to GBA	Cross loading with other variable
	S2	I would feel comfortable when using accessible facilities in my travel to GBA	Cross loading with other variables

Table 5 Result of Exploratory Factor Analysis

Variabiles	1	2	Component 3	4	5	6	Mean	Std. Deviation	Variance Extracted Explained(%)	KMO
A2	–	0.79	–	–	–	–	5.08	1.216	17.770	0.89
A3	–	0.81	–	–	–	–	5.12	1.130	–	–
A4	–	0.63	–	–	–	–	5.44	0.904	–	–
A5	–	0.56	–	–	–	–	5.51	0.997	–	–
A6	–	0.67	–	–	–	–	5.33	0.967	–	–
A7	–	0.80	–	–	–	–	4.84	1.189	–	–
U6	–	0.59	–	–	–	–	5.11	1.192	–	–
U1	–	–	–	0.82	–	–	5.42	1.120	11.947	0.82
U2	–	–	–	0.74	–	–	5.50	0.990	–	–
U3	–	–	–	0.77	–	–	5.48	1.044	–	–
U4	–	–	–	0.65	–	–	5.54	0.918	–	–
S3	–	–	–	–	–	0.78	5.72	0.917	4.899	0.50
S4	–	–	–	–	–	0.60	5.46	0.964	–	–
PHI	–	–	0.88	–	–	–	5.06	1.399	16.364	0.88
PH2	–	–	0.83	–	–	–	4.91	1.498	–	–
PH3	–	–	0.92	–	–	–	5.29	1.343	–	–
PH4	–	–	0.93	–	–	–	5.27	1.364	–	–
PH5	–	–	0.92	–	–	–	5.28	1.397	–	–
TS1	–	–	–	–	0.70	–	5.62	0.963	9.884	0.87
TS2	–	–	–	–	0.69	–	5.55	0.936	–	–
TS3	–	–	–	–	0.70	–	5.59	0.931	–	–
TS4	–	–	–	–	0.69	–	5.55	0.956	–	–
Til	0.83	–	–	–	–	–	5.54	0.999	19.693	0.89
TI2	0.86	–	–	–	–	–	5.47	1.083	–	–
TI3	0.86	–	–	–	–	–	5.57	0.980	–	–
TI4	0.88	–	–	–	–	–	5.49	1.054	–	–
TI5	0.85	–	–	–	–	–	5.55	1.022	–	–
Total		–	–	–	–	–	–	–	80.56	0.94

Extraction Method: Principal Component Analysis
[a]Rotation converged in 6 iterations

the heterotrait-monotrait (HTMT) ratio of correlations. Henseler et al. [48] demonstrated the superior performance of the method through a Monte Carlo simulation study.

According to HTMT criterion, if the value of HTMT was below HTMT.90 value of 0.90, the discriminant validity between the two constructs had been established.

Table 6 Construct reliability and validity

Constructs	Items	Loadings	CR	AVE	Cronbach's alpha
Accessibility	A2	0.66	0.91	0.6	0.9
	A3	0.79			
	A4	0.83			
	A5	0.87			
	A6	0.83			
	A7	0.72			
	U6	0.69			
Usability	U1	0.79	0.93	0.77	0.9
	U2	0.9			
	U3	0.91			
	U4	0.91			
Security	S3	0.9	0.88	0.78	0.72
	S4	0.86			
Perceived helplessness (PH)	PH1	0.92	0.96	0.84	0.95
	PH2	0.83			
	PH3	0.94			
	PH4	0.95			
	PH5	0.94			
Travel satisfaction (TS)	TS1	0.94	0.97	0.9	0.96
	TS2	0.94			
	TS3	0.97			
	TS4	0.96			
Revisit Intention (RI)	RI1	0.95	0.98	0.89	0.97
	RI2	0.92			
	RI3	0.96			
	RI4	0.94			
	RI5	0.96			

Table 7 Fornell–Larcker criterion

Variables	1	2	3	4	5	6
Accessibility	0.77					
Usability	0.66	II.SH				
Security	−0.23	0.63	0.88			
Perceived Helplessness	0.57	0.59	0.62	0.92		
Travel Satisfaction	0.77	0.49	0.55	−0.35	0.95	
Revisit Intention	0.76	0.88	0.88	−0.36	0.77	0.94

As shown in Table 8, the value of each construct was below 0.9. Hence, the variables accessibility, usability, security, PH, TS and RI were valid.

Under the premise of ensuring the discriminant validity, the adjustment of the measurement model had been completed. It ensured the reliability of the convergence and discriminant validity of the structural equation model (SEM). Hair et al. [43] proposed looking at R^2, Beta, and the related t-values by resampling the bootstrapping method of 5000 resampling to evaluate the structural model. This process was taken place in SMART PLS 3.3.3 software.

The R^2 evaluated partial variances of endogenous variables, which were explained by the structural model. This indicated the quality of the adjusted model. In the field of social and behavioral science, Cohen [49] suggested that $R^2 = 2\%$ as classified as having a small effect, $R^2 = 13\%$ as a medium effect, and $R^2 = 26\%$ as having a large effect. Table 9 showed the values of R^2. Results indicated the latent variables travel satisfaction (TS) and revisit intention (RI) had medium effect (12%), and for the latent variables perceived helplessness (PH), the R2 was quite small (7%).

Then relationships between variables was tested. Table 10, Table 11 and Fig. 3 presented the path coefficients and the result of hypotheses test. Literature stated that for a correlation case, a t-value greater than 1.96 was considered to show a significant relationship, while the p-value should be less than or equal to 0.05. H1 and H2 were rejected but the remaining hypotheses were supported since their t-values and p-values were all within acceptable limits.

What was noteworthy was that the path coefficients of H3, H4a, H4b and H5 were respectively −0.17, −0.35, −0.10 and 0.74, indicating the security of the accessible environment negatively affected tourists' level of perceived helplessness, and the decrease of tourists' level of perceived helplessness would effectively improve tourists' travel satisfaction and their intention to revisit. The improvement of tourists' travel satisfaction significantly increased their revisit intention.

Table 8 Discriminant validity: Heterotrait-monotrait (HTMT)

Constructs	1	2	3	4	5 6
Accessibility					
Usability	0.808				
Security	0.805	0.778			
Perceived helplessness	0.177	0.235	0.303		
Travel satisfaction	0.684	0.622	0.751	0.36	
Revisit intention	0.593	0.518	0.667	0.367	0.796

Table 9 R^2 valves of dependent variables

Constructs	R^2
Perceived helplessness (PH)	0.07
Travel satisfaction (TS)	0.12
Revisit intention	0.12

Table 10 t-Values, p-value and path coefficients between variables

Path	Hypotheses	t-Values	p-Values	Decision
A→PH	HI	0.78	0.22	Rejected
U→PH	H2	0.88	0.19	Rejected
S→PH	H3	2.54	0	Supported
PH→TS	H4a	6	0	Supported
PH→RI	H4b	1.99	0.03	Supported
TS→RI	H5	9.71	0	Supported

Table 11 Results of hypotheses test

Path	t Statistics	p Values	Path Coefficients
A→PH	0.78	0.22	−0.07
U→PH	0.88	0.19	−0.06
S→PH	2.54	0	−0.17
PH→TS	6	0	−0.35
PH→RI	1.99	0.03	−0.1
TS→RI	9.71	0	0.74

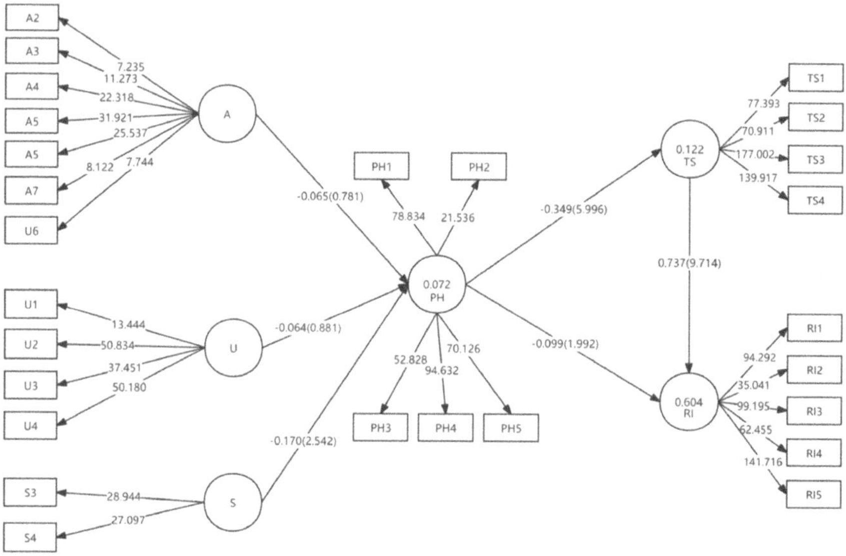

Fig. 3 Path coefficients and t-values among constructs

Table 12 f^2 Values and Q^2 values of each variable

Constructs	f^2	Q^2
Accessibility	\	0.47
Usability	\	0.6
Security	\	0.31
Perceived helplessness	0.07	0.75
Travel satisfaction	0.14	0.81
Revisit intention	0.14	0.82

Two other indicators of model adjustment need to be evaluated, which were relevance or predictive validity (Q^2) (Stone–Geisser indicator), and effect size (f^2)(Cohen's indicator). Hair et al. [43] stated the Stone–Geisser indicator (Q^2) assessed the degree to which the model was close to expectations. As a criterion for evaluation, a value greater than zero should be obtained. Cohen's indicator (f^2) was obtained by including and excluding model constructs one by one. The value of each construct to the adjustment model would be assessed. 0.02 was considered to have a small effect, while 0.15 was medium and 0.35 is large.

The interpretation of Table 12 showed that the values of Q^2, as well as f^2. The Q^2 value of each construct was higher than 0, among them, Q^2 value of perceived helplessness (PH), travel satisfaction (TS) and revisit intention (RI) were ranged from 0.31 to 0.82, which was close to 1, showing that the model reflected the actual situation to a great extent. Even though the f^2 value of perceived helplessness was not high, the other two f^2 values of travel satisfaction and revisit intention were closing reach to 0.15, showing medium effect to the model. In general, all values were within the recommended range, indicating the model was accurate and the constructs were crucial to the model.

The independent sample t-test was used to examine whether there were mean differences between person with impaired physical mobility and person without impaired physical mobility in each dimension.

Table 13 indicated that person with impaired physical mobility and person without impaired physical mobility had significant mean differences in the dimensions of accessibility, usability, security, and perceived helplessness (PH) (p value ≤ 0.05).

The independent sample t-test was conducted again to further explore which items were different between people with impaired physical mobility and people without impaired physical mobility. Results were shown in Table 14.

Person with impaired physical mobility and person without impaired physical mobility had significant mean differences between the items A4, A5, A6 and A7 from accessibility, U1, U2, U3 and U4 from usability, S3 from security, PH1, PH2, PH3, PH4 and PH5 from perceived helplessness (PH). It could be seen that the accessibility of transportation modes, accessible facilities, information and services in soft environments in the Greater Bay Area have a significant impact on tourists with impaired physical mobility.

Table 13 Mean differences between person with impaired physical mobility and person without impaired physical mobility

Variables	People with impaired physical mobility	P$eople without impaired physical mobility	T value	2-Tail Sig
Accessibility	5.28	5.1	2.16	0.03
Usability	5.72	5.15	6.86	0
Security	5.72	5.21	5,82	0
Perceived Helplessness	5.44	4.75	5.58	0
Travel satisfaction	5.6	5.54	0.66	0.51
Revisit intention	5.47	5.61	−1.41	0.16

Table 14 Independent sample t test for each item

Variables	Items	People with impaired physical mobility	People without impaired physical mobility	T value	2-tail Sig
Accessibility	A2	5.11	5.05	0.52	0.6
	A3	5.13	5.11	0.17	0.H7
	A4	5.54	5.29	2.85	0
	A5	5.63	5.33	3.03	0
	A6	5.42	5.2	2.25	0.02
	A7	4.97	466	2.65	0 01
	U6	5.16	5.04	1.04	0.3
Usability	U1	5.7	5.02	2.65	0.01
	U2	5.75	5.15	6.4	0
	U3	5.74	5.11	6.33	0
	U4	5.71	5.3	4.53	0
Security	S3	5.8	5.6	2.2	0.03
	S4	5.43	5.5	−0.74	0.46
Perceived helplessness	PHI	5.39	4.58	6.06	0
	PH2	5.36	4.25	8.01	0
	PH3	5.46	5.04	3.21	0
	PH4	5.51	4.92	4.44	0
	PH5	5.49	4.98	3.66	0

Meanwhile, tourists valued the universality, flexibility and comfort of usability in an accessible environment. Compared with tourists without impaired physical mobility, tourists with impaired physical mobility were more sensitive to violence or intimidation during travel due to differences in physical functions and the level of perceived helplessness of them was obviously higher than travelers without impaired physical mobility.

5 Conclusion and Discussion

The paper investigated the impact of accessible environment on tourists' travel satisfaction and revisit intention in Greater Bay Area. Considering the purpose and objectives of this study, six hypotheses were tested, of which four were supported and the other two were rejected. The first main finding corroborated that security of accessible environment, as expected, significantly affected tourists' perceived level of helplessness related to travel ($\beta = -0.17$, $p < 0.05$). This result was aligned with findings of Kiani, Khodabakhsh [50]. They showed that effectiveness of security training significantly reduced helplessness. With the consideration of security issues and the use of protection devices decreased the injury chance of tourists in the journey.

The result revealed that with the decrease of the perceived level of helplessness, tourists' travel satisfaction would significantly increase ($\beta = -0.35$, $p < 0.05$). The third main conclusion that could be drawn was that tourists' perceived level of helplessness in the Greater Bay Area, had statistically significant impact on tourists' revisit intention ($\beta = -0.36$, $p < 0.05$). This finding confirmed tourists' revisit intention would decrease when their perceive level of helplessness became more and more grievous [24].

Furthermore, the study confirmed that compared with people without impaired physical mobility, tourists with impaired physical mobility would pay more attention to the accessibility, usability and security of accessible environment and be more likely to perceived helplessness. This was determined by the capacity limitations of tourists with impaired mobility, which was consistent with previous research [51].

In addition, the findings demonstrated travel satisfaction positively affected tourists' revisit intention and further confirmed the results of previous studies [29, 30, 35].

However, the results revealed that the effect of accessibility and usability of accessible environment on tourists' perceived level of helplessness were insignificant. This reflected that the basic accessibility and usability requirements could not meet the tourists' expectations for an accessible tourism environment. Many accessibility facilities had been built without considering the actual needs of people, just to meet regulatory pressures. Once the relevant departments loosen regulation, the guarantee of accessibility and usability would be difficult to maintain.

Academically, this study contributed to the current hospitality literature to emphasize the accessible environment dimension and its relationship with tourists' travel satisfaction and revisit intention. This study further verified the application of the theory of learning helplessness in the field of tourism research. In this study, the study population was expanded from the elderly, the disabled and other vulnerable groups to ordinary people, indicating that even ordinary people would feel helpless in an unfamiliar environment.

In terms of practice, the findings of this study revealed the importance of security of accessible environment and tourists' perceived level of helplessness. Since the security of accessible environment effectively decreased tourists' perceived level of helplessness, local authorities in the Greater Bay Area could improve the operating comfort and safety of accessible facilities, providing a relaxed and safe experience for tourists. In addition, leaders of hotels, shopping malls and other public operating places in the Greater Bay Area were suggested to pay attention to staff training and help staff timely found and dealt with unexpected situations encountered by tourists.

6 Limitations and Future Research

This research, like any others, has certain limitations that allow for further investigation.

Firstly, the development of urban accessible environment in Greater Bay Area is at different levels, and study the Greater Bay Area as a whole might lead to bias in final results. In future research, cities with subways in the Greater Bay Area could be further selected as research destinations, because subways are the preferred means of transportation for people with impaired physical mobility.

Secondly, the accessible environment in this study is considered as a secondary structure composed of three dimensions (accessibility, usability and security). However, Morin et al. [52] as well as Kim, Moon [53] have proposed in their studies on the physical environment that the perception of the physical environment could also be considered as a whole. Furthermore, they suggest an individual's reaction to the environment is determined by the incorporation of multisensory information. As a result, further research could broaden this study to include more comprehensive accessibility research.

References

1. Open Doors Organization: Research among adults with disabilities—travel and hospitality. Open Doors Organization, Chicago (2002)
2. Sellick, M.C.: Discovery, connection, nostalgia: key travel motives within the senior market. J. Travel Tour. Mark. **17**(1), 55–71 (2004)
3. Wennberg, H., Hydén, C., Ståhl, A.: Barrier-free outdoor environments: older peoples' perceptions before and after implementation of legislative directives. Transp. Policy **17**(6), 464–474 (2010)
4. Wu, W.D., Chang, C.F., Lu, T.H., Tsai, C.J.: The study of accessible campus environment from the viewpoint of students with disabilities. Bull. Spec. Educ. **7**, 23–41 (1991)
5. Masood, H., Shaheen, S.S.: Barrier free environment: an analysis of Aligarh City. India. Int. J. Interdiscip. Multidiscip. Stud. **1**(9), 8–15 (2014)
6. Jiao, J., Sun, L., Yang, M.: Accessible design in the city. China Architecture Building Press, Beijing (2014)
7. Walls, A.R., Okumus, F., Wang, Y.R., Kwun, D.J.W.: An epistemological view of consumer experiences. Int. J. Hosp. Manag. **30**(1), 10–21 (2011)

8. Hansen, W.G.: How accessibility shapes land use. J. Am. Plan. Assoc. **25**(2), 73–76 (1959)
9. Dalvi, M.Q., Martin, K.M.: The measurement of accessibility: some preliminary results. Transportation **5**(1), 17–42 (1976)
10. Burns, L.D.: Transportation, temporal, and spatial components of accessibility. Lexington Books, Toronto (1979)
11. Henry, S.L., Abou-Zahra, S., Brewer, J.: The role of accessibility in a universal web. Paper presented at the Proceedings of the 11th Web for All Conference, Seoul, Korea, 7 April (2014)
12. Geurs, K.T., van Wee, B.: Accessibility evaluation of land-use and transport strategies: review and research directions. J. Transp. Geogr. **12**(2), 127–140 (2004)
13. Lättman, K., Friman, M., Olsson, L.E.: Perceived accessibility of public transport as a potential indicator of social inclusion. Soc. Incl. **4**(3), 36–45 (2016)
14. Rumetshofer, H., Wöß, W.: Tourism information systems promoting barrier-free tourism for people with disabilities. In: Miesenberger, K., Klaus, J., Zagler, W.L., Burger, D. (eds.) Computers helping people with special needs, pp. 280–286. Springer, Berlin (2004)
15. International Standards Organization: ISO 9241-11: Ergonomic requirements for office work with visual display terminals (VDTs). Part 11: Guidance on usability. International Standards Organization, Geneva (1998)
16. Petrie, H., Bevan, N.: The evaluation of accessibility, usability, and user experience. In: Stepanidis, C. (ed) The universal access handbook. CRC Press, Boca Raton, pp 20–21–20–16 (2009)
17. Wegge, K.P., Zimmermann, D.: Accessibility, usability, safety, ergonomics: Concepts, models, and differences. In: Stephanidis, C. (ed.) Universal acess in human computer interaction, pp. 294–301. Coping with diversity. Springer, Berlin (2007)
18. Dattolo, A., Luccio, F.F., Pirone, E.: Webpage accessibility and usability for autistic users: a case study on a tourism website. In: Paper presented at the ninth international conference on advances in computer–human interactions (ACHI 2016), Venice, 24–28 April (2016)
19. Zhang, Y., Woo Park, Y., Cole, S.T.: Factors influential to tourism experiences of people with mobility impairments. TTRA. (2016)
20. Mehta, V.: Evaluating public space. J. Urban Des. **19**(1), 53–88 (2014)
21. Beecroft, M.: The future security of travel by public transport: a review of evidence. Res. Transp. Bus. Manag. **32**, 100388 (2019)
22. Beecroft, M., Pangbourne, K.: Future prospects for personal security in travel by public transport. Transp. Plan. Technol. **38**(1), 131–148 (2015)
23. Miller, W.R., Seligman, M.E.: Depression and learned helplessness in man. J. Abnorm. Psychol. **84**(3), 228–238 (1975)
24. Lee, B.K., Agarwal, S., Kim, H.J.: Influences of travel constraints on the people with disabilities' intention to travel: an application of seligman's helplessness theory. Tour. Manag. **33**(3), 569–579 (2012)
25. Wen, J., Huang, S., Goh, E.: Effects of perceived constraints and negotiation on learned helplessness: a study of Chinese senior outbound tourists. Tour. Manag. **78**, 104059 (2020)
26. Wen, J.: Do parents enjoy travelling with their young children? An application of learned helplessness theory. Anatolia **31**(3), 511–513 (2020)
27. Specht, J., King, G., Brown, E., Foris, C.: The importance of leisure in the lives of persons with congenital physical disabilities. Am. J. Occup. Ther. **56**(4), 436–445 (2002)
28. Ying, T., Tang, J., Wen, J., Ye, S., Zhou, Y., Li, F.: Traveling with pets: constraints, negotiation, and learned helplessness. Tour. Manag. **82**, 104183 (2021)
29. An, M., Lee, C., Noh, Y.: Risk factors at the travel destination: their impact on air travel satisfaction and repurchase intention. Serv. Bus. **4**(2), 155–166 (2010)
30. Rittichainuwat, B.N., Qu, H., Mongknonvanit, C.: A study of the impact of travel satisfaction on the likelihood of travelers to revisit thailand. J. Travel Tour. Mark. **12**(2–3), 19–43 (2015)
31. Guntoro, B., Hui, T.K.: Travel satisfaction and revisit intention of chinese visitors: The case of Singapore. Adv. Hosp. Leis. **9**, 29–47 (2013)
32. Hui, T.K., Wan, D., Ho, A.: Tourists' satisfaction, recommendation and revisiting Singapore. Tour. Manag. **28**(4), 965–975 (2007)

33. Hui, T.K., Wan, D., Cheng, H.Y.: Case study of tourist revisit behavior in Singapore. Adv. Hosp. Leis. **6**, 27–42 (2010)
34. Gyte, D.M., Phelps, A.: Patterns of destination repeat business: British tourists in Mallorca. Spain. J. Travel Res. **28**(1), 24–28 (1989)
35. Lau, A.L.S., McKercher, B.: Exploration versus acquisition: a comparison of first-time and repeat visitors. J. Travel Res. **42**(3), 279–285 (2004)
36. Campo-Martínez, S., Garau-Vadell, J.B., Martínez-Ruiz, M.P.: Factors influencing repeat visits to a destination: the influence of group composition. Tour. Manag. **31**(6), 862–870 (2010)
37. Lam, T., Hsu, C.H.C.: Predicting behavioral intention of choosing a travel destination. Tour. Manag. **27**(4), 589–599 (2006)
38. Zarrad, H., Debabi, M.: Analyzing the effect of electronic word of mouth on tourists' attitude toward destination and travel intention. Int. Res. J. Soc. Sci. **4**(4), 53–60 (2015)
39. Leong, L.Y., Hew, T.S., Lee, V.H., Ooi, K.B.: An SEM-artificial-neural-network analysis of the relationships between SERVPERF, customer satisfaction and loyalty among low-cost and full-service airline. Expert Syst. Appl. **42**(19), 6620–6634 (2015)
40. Chen, F., Dai, S., Xu, H., Abliz, A.: Senior's travel constraint, negotiation strategy and travel intention: examining the role of social support. Int. J. Tour. Res. **23**(3), 363–377 (2021)
41. United Nations: Declaration on the rights of disabled persons. UN. (1975)
42. Jia, A., J.: Barrier free design. Chemical Industry Press, Beijing (2015)
43. Hair, J.F., Hult, G.T.M., Ringle, C., Sarstedt, M.: A primer on partial least squares structural equation modelling (PLS-SEM). Sage Publications, Los Angeles (2013)
44. Hair, J.F., Black, W., Babin, B.: Multivariate data analysis. Pearson, New York (2009)
45. Kaiser, H.F.: An index of factorial simplicity. Psychometrika **39**(1), 31–36 (1974)
46. Fornell, C., Larcker, D.F.: Evaluating structural equation models with unobservable variables and measurement error. J. Mark. Res. **18**(1), 39–50 (1981)
47. Bido, D.D.S., Da Silva, D., Ringle, C.M.: Structural equation modeling with the SmartPLS. Brazilian J. Mark. **13**(2), 56–73 (2014)
48. Henseler, J., Ringle, C.M., Sarstedt, M.: A new criterion for assessing discriminant validity in variance-based structural equation modeling. J. Acad. Mark. Sci. **43**(1), 115–135 (2015)
49. Cohen, J.: Statistical power analysis for the behavioral sciences. Routledge, New York (2013)
50. Kiani, F., Khodabakhsh, M.R.: Promoting individual learning for trainees with perceived high helplessness: experiences of a safety training program. Iran. J. psychiatry Behav. Sci. **8**(4), 19–28 (2014)
51. Petrie, H., Kheir, O.: The relationship between accessibility and usability of websites. In: Paper presented at the Proceedings of the SIGCHI Conference on Human Factors in Computing Systems, California, USA, 28 April-3 May (2007)
52. Morin, S., Dubé, L., Chebat, J.C.: The role of pleasant music in servicescapes: a test of the dual model of environmental perception. J. Retail. **83**(1), 115–130 (2007)
53. Kim, W.G., Moon, Y.J.: Customers' cognitive, emotional, and actionable response to the servicescape: a test of the moderating effect of the restaurant type. Int. J. Hosp. Manag. **28**(1), 144–156 (2009)

The Perceived Value of Local Knowledge Tourism: Dimension Identification and Scale Development

Hailin Zhang, Jinbo Jiang, and Jinsheng Zhu

Abstract Local knowledge performs a unique function in emphasizing the locality, originality, and mythology of a destination, but its tourist perceived value entailed further exploration. Through a constructivist grounded theory approach, this study utilize Guilin of China as a case to probe into the tourist perceived value of local knowledge. 22 respondents were solicited for their local knowledge tourism experience and perception while related secondary data extracted from major Chinese online travel agencies. The thematic analysis has identified 19 concepts and 5 dimensions of the tourism perceived values of local knowledge, including the Function Value (FV), Emotional Value (EV), Social Value (SV), Cognition Value (CV) and Self-actualization Value (SAV). Based on this a scale with 19 items has been developed and tested following the rigorous procedure. The study expands the research context of perceived value in tourism and provides a theoretical insights on product development for local knowledge tourism.

Keywords Local knowledge tourism · Perceived value · Value dimension · Scale development

1 Introduction

Local knowledge is the traditional knowledge, values, skills, beliefs, and philosophy that civilizations and communities have created through long-term contact with their natural and humanistic environments. Local knowledge has substantial social,

H. Zhang · J. Jiang
South China University of Technology, Guangzhou, Guangdong, People's Republic of China
e-mail: 308969979@qq.com

J. Jiang
e-mail: 1830429094@qq.com

J. Zhu (✉)
Guilin Tourism University, Guilin, Guangxi, People's Republic of China
e-mail: 328031240@qq.com

Y. Luo et al. (eds.), *Tourist Experience and Perception in the Pandemic and Post-Pandemic Era*, https://doi.org/10.1007/978-981-96-1820-0_3

cultural, ecological, scientific, and economic importance and shows the variety of human civilization and culture [1]. This knowledge has been formed through time and transferred in a multitude of settings, including via song, dance, sculptures, and rituals [2]. In the field of tourism, local knowledge is acknowledged as a significant tourist attraction and has unique value for cultural and tourism development. In order to address sustainability and social justice in the tourist industry, it is essential to recognize the diversity of cultures and the inherent multifaceted modalities of knowledge [3, 4]. Local knowledge tourism is developed with an emphasis on the elements of knowledge in heritage tourism and significance of facilitating the dialogue and exchange between diverse cultures. As such, it is a constructive reaction to the requirement of deepening and strengthening tourist experience in the era of culture and tourism integration. Unlike heritage tourism that merely promotes superficial historical landscape viewing and experience, local knowledge tourism stresses in-depth visitor involvement and knowledge development. Existing research has not given sufficient attention to visitors' subjective impression and experience of an attraction's worth, and the structure, features, and assessment of relevant experience value must be elucidated. Thus, the present academic research leaves a substantial research gap in further comprehending the perceived value of local knowledge tourism, necessitating brand-new grounded research and interpretation based on a solid theoretical foundation.

The definition of tourist perceived value is derived from the marketing concept of customer values and was reintroduced into tourism studies in the 1990s [5]. Since the dawn of the twenty-first century, the concept has been extended to tourist consumer behaviour and destination marketing research. It has also emerged as a new prominent research topic, following in the footsteps of quality management and visitor satisfaction [6]. This notion offers the perfect vantage point from which to comprehend the tourists' all-encompassing assessments of their travel experiences within the framework of the consumption-based model of local knowledge tourism. This study will expands the research context of perceived value in tourism and deepens the understanding of local knowledge tourism regarding tourist experience and perception. It will also contribute to a deeper comprehension of tourist behavioural characteristics, consumption behaviour and psychology, thereby furnishing a theoretical foundation for the development of products and a marketing strategy for local knowledge tourism. For these reasons, the current investigation expects to utilize Guilin as a research case study. Guilin, located in the Guangxi Zhuang Autonomous region of China and to the northeast of the region's capital city, is a well-known international tourist destination that is renowned for its beautiful scenery and unique indigenous culture. This exploratory research applies the grounded theory in order to determine the multi-dimensional structure of the value that tourists might perceive in local knowledge tourism. In addition to this, a local knowledge tourism perceived value scale is also developed in accordance with the stringent scientific procedure. This procedure includes item development, item purification, questionnaire surveying, exploratory factor analysis, and confirmatory factor analysis, amongst other issues.

This research seeks to give a theoretical support and practical direction for the development of local knowledge tourism, as well as fill a gap in the existing body of literature.

2 Literature Review

2.1 Local Knowledge and Local Knowledge Tourism

Clifford Geertz, an American anthropologist, is attributed with introducing the idea of "local knowledge" as a central notion in interpretative anthropology; nevertheless [7], he did not provide a precise term for the concept. Local knowledge has been characterized in a variety of ways, depending on the setting and the goals of various researchers and organizations. For instance, the United Nations' Local and Indigenous Knowledge Systems Program (LINKS), which aims to support the preservation and transmission of collective memory and cultural heritage, defines that understandings, capacities, and philosophies formed by communities with extensive histories of engagement with their natural environs are referred to as local knowledge [8]. Local knowledge plays an essential role in a cultural network that also includes linguistic systems, categorization schemes, techniques for making use of resources, social interactions, and rituals and spiritual activities. Local knowledge, on the other hand, is a form of unofficial and peripheral knowledge held by indigenous people. This stands in contrast to the universal scientific knowledge that is centered on the west. Broadly speaking, it is the sum of material and cultural achievements accumulated by people in a certain region during their historical development, involving every aspects of life that may include local economic development, science and technology, social values, religious beliefs, culture and arts, social customs, lifestyles and social codes of conduct [9, 10]. It is not always related to specific daily life, but also contains abstract thoughts, philosophies and insights, which are comprehensive knowledge and technology about ecology, geography or society that revitalize the locale [11]. Such knowledge can highlight the diversity and equality of different cultures. The mysterious local knowledge is inherently attractive to the general public, by which such knowledge is related to the cultural economy [12], festivals creation [13], co-production of histories [14], as well as ethnic tourism development issues [15]. Integrating local knowledge into the design of participatory tourism can highly promote the quality of tourism development, participation of local community in tourism planning and operation, as well as heritage preservation [16–18]. Deep experience of the local knowledge at a destination, for instance, the local engagements [19] and indigenous tourism experiences [20] will contribute to the tourist experience and lead to in-depth tourism. A consensus has been reached that enhancing cultural integration and tourist experience in destination management has become a significant trend for tourism planning and products development [21–23]. Based on the above discussion, this study defined local knowledge tourism as a type of heritage tourism that utilize local knowledge of a destination as its essence of tourism attraction, aiming to conserve the traditional cultural heritage and provide an in-depth and participatory cultural tourism experience and knowledge acquisition opportunity to tourists. From the perspective of tourist, one of the major goals or motivations is to gain local indigenous knowledge of the destination. Since local knowledge encompasses the knowledge and practices contextualised in the local daily life experiences, such tourist activities can offer a deeper cultural experience compared to other types of heritage tourism.

2.2 *Tourism Perceived Value*

2.2.1 The Definition of Tourism Perceived Value

Customer perceived value was first discussed in the field of marketing and refers to how customers feel about a product or service with regards to perceived quality, internal and external features of the product and other psychological benefits [24]. It has been defined as a customers' assessment of the trade-offs between the benefits and sacrifices realized in selecting a given product from the options available in the market [25–27]. Perceived costs are mainly composed of monetary costs and non-monetary costs which relates to the costs of time, physical efforts or life style changes that one has to pay to acquire the product or service [28]. As a special form of customer perceived value, tourism perceived value is the emotional and psychological benefits generated by a series of interaction between tourists and the outside world. It is an important research dimension to grasp tourists' perception and experience of tourist destinations [29]. The definition and interpretations of tourist perceived value vary considerably [30, 31], but there is a consensus on the core connotation, namely that it is the trade-off between perceived gains and losses of tourists in a specific tourism context and it is an overall evaluation of tourist products or services. Perceived value is deemed to have great impact on tourist preference, satisfaction and loyalty, for example, high level of perceived value will lead to the repurchase of the products and service [32]. Therefore, measuring the perceived value of tourists may contribute to a better understanding of tourists' consumption behaviour and psychology, as well as provide theoretical basis for tourism marketing [33]. The circumstances, quality and attributes of tourism products have a significant role in determining the tourist perceived value. Meanwhile, it is also influenced by subjective elements such as preference, attitude, physiological condition, demand, and motivation [34, 35]. Compared to consuming activities in other commercial sectors, tourist consumption is more complicated and diverse, and the perception of tourism's worth has plural and structural dimensions. The perceived value of a tourist destination contains functional value, hedonic value and symbolic worth. Due to the intangible, diachronic and interactive features of tourism products and services, tourists tend to perceive the value of tourism experiences from a holistic perspective considering multiple elements such as a product, service and destination [36].

2.2.2 Dimensions and Measurement of Tourist Perceived Value

Customers perceived value are mainly assessed in the unidimensional approach [27] and the multidimensional approach (Sheth et al., 1991). The unidimensional approach emphasizes economic utility apart from the trade-offs between the benefits such as customer utility and sacrifices such as price, time, effort etc. [37, 38]. A typical practice is to incorporate the tourist perceived value as an independent variable into the model of tourist satisfaction or loyalty [30]. To overcome the limitations of the unidimensional approach that ignores various aspects of a person's emotional state and external conditions, the multidimensional approach was proposed to explore the factors underlying the phenomenon [38, 39]. Using the multidimensional approach, Sheth et al. [38] identified five dimensions of customer perceived value that include social, emotional, functional epistemic and conditional value. Sweeney and Soutar (2001) [39] synthesize previous studies [40–42] to develop the four-dimension PERVAL scale in which emotional value represents the utility derived from the feelings or affective states that a product generates; social value refers to the utility derived from the product's ability to enhance social relationship; functional value (price/value for money) which represents the utility derived from the product due to the reduction of its perceived short-term and long-term costs; and functional value (performance/quality) which refers to the utility derived from the perceived quality and expected performance of the product [39].

To identify and assess the tourist perceived values, scales have been modified and developed in different tourism contexts such as the five-dimensional SERV-PERVAL scale that encompasses quality, monetary price, non-monetary price, reputation and emotional response for cruise tourism [43]. Andrades and Dimanche (2018) analyse tourist value perception from the perspective of psychology and developed a scale of co-creating tourists' experiential value at tourist attractions [44]. In addition, related scales in tourism and hospitality research may involve the overall perception of a tourist destination, snow and ice tourism, ancient village tourism and wetland park tourism as illustrated in Table 1. These studies might expand our understanding of consumer value in the context of tourism and serve as a point of reference for the current inquiry. In addition to providing a reliable tool for quantitative study in the context of local knowledge tourism, the following new scale investigation may also fill the gap in the literature on tourist experience.

3 Research Design

3.1 Research Method and Case Selection

Due to the exploratory nature of this research, a constructivist grounded theory and content analysis are applied to analyse the data collected until theoretical saturation. Grounded theory is useful to develop theory based on the analysis of systematically

Table 1 Examples of scale development of tourist perceived value

Authors and publication year	Scenarios	Measurement of perceived value
Sheth et al. [38]	Commodities	5 dimensions: functional value, social value, emotional value, conditional value, and epistemic value
Ling & Jigang [45]	Traditional event tourism	7 dimensions: Cultural epistemic value, hedonic value, social value, price and convenience value, service value, conditional value and overall value, Environmental value, functional value
Petrick [43]	Cruise tourism	5 dimensions: quality value, emotional value, cost value, behavior cost value and reputation value
Mengfan et al. [46]	Tourist destination	3 core dimensions: functional value, hedonic value and symbolic value, and 8 sub-categories: relaxation, healing, communication and social prestige
Pi et al. [47]	Tourist destination	4 dimensions: functional value, social value, emotional value and brand value
Li et al. [28]	Living Transmission of intangible cultural heritage	5 dimensions: inheritance value, context value, living state value, economic value and cognitive value

collected data and through interaction with study participants [48]. To realize the ultimate goal of theoretical construction with high reliability and validity, further to the primary data from the semi-structured interviews, secondary data are extracted from sources like online travelogue and news report to form a triangle cross validation. In the text encoding and follow-up study, review and advice are also sought from a group of experts for item purification. In the verification part, SPSS26.0 and AMOS24.0 were used for confirmatory factor analysis, reliability and validity analysis, and competitive factor analysis.

Guilin city in China is chosen as the study case because it is a world-famous tourist destination and a recognized national historical and cultural city, as designated by the Chinese central government. Additionally, Guilin is home to a dozen of ethnic minorities, including the Zhuang, Dong, Miao, and Yao. With extensive natural, cultural, and tourist resources, Guilin tourism contains rich local knowledge in a variety of forms, from traditional agricultural lifestyle to genuine ethnic culture, making it an appropriate case study for this research.

3.2 Data Collection

Without establishing the structure of the perceived value prior to the grounded theory analysis, the first step is to solicit the views of the respondents regarding four aspects of their local knowledge tourism experience: tourism motivation, experience process, experience value perception, personal feelings and gain. From March

Table 2 Profile of the respondents (N = 22)

No	Gender	Age	Occupation	Origin	No	Gender	Age	Occupation	Origin
T1	F	21–30	Student	Guangxi	T2	M	21–30	Student	Jiangsu
T3	M	41–50	Company employee	Beijing	T4	F	21–30	Company employee	Guangxi
T5	F	31–40	Teacher	Guizhou	T6	F	21–30	Teacher	Guangxi
T7	M	31–40	Teacher	Guizhou	T8	F	31–40	Researcher	Guangxi
T9	F	41–50	Self-employed	Guangxi	T10	M	31–40	Company employee	Guangxi
T11	F	51–60	Company employee	Guangxi	T12	M	31–40	Company employee	Guangxi
T13	F	21–30	Self-employed	Guangxi	T14	M	51–60	Retiree	Guangxi
T15	M	21–30	Student	Hunan	T16	F	21–30	Student	Guangxi
T17	M	21–30	Self-employed	Guangxi	T18	F	21–30	Student	Guangxi
T19	F	41–50	Company employee	Guangxi	T20	M	21–30	Company employee	Guangxi
T21	F	31–40	Self-employed	Shanghai	T22	F	31–40	Doctor	Guangxi

to June 2022, semi-structure interviews were performed in five local knowledge tourism experience zones in Guilin: East West Alley, Terrace Field of Longsheng, Guangxi Ethnic Tourism Museum, West Street of Yangshuo and venue of Impression Show of Liusanjie. The interview asks respondents to discuss and react the following personal experience-based questions. To begin, could you please describe the steps involved in your local knowledge tour? Which aspects of the local culture did you get to encounter while you were there? What do you consider to be the benefits of having local knowledge while participating in tourist activities? What unique experiences and personal advantages have you gained as a direct result of the touring experience? Each interview lasted about 20–30 min, and the whole process was voice recorded with the permission of the interview participants. According to the selection criteria the participants needed to have conducted a local knowledge tourism in Guilin as defined by this study to ensure the integrity of the experience. 22 participants were recruited through snowball sampling and their demographic profile are shown in Table 2.

3.3 *Coding and Perceived Values Identification*

Using a method known as thematic analysis, the data were then evaluated and examined. Using the professional qualitative research software Nvivo 11.0 for data coding, the thematic analysis has identified 19 concepts and 5 dimension of the the tourism perceived values of local knowledge, including the Function Value (FV), Emotional

Value (EV), Social Value (SV), Cognition Value (CV) and Self-actualization Value (SAV) as shown in Tables 3 and 4.

3.4 Scale Development

The initial scale items of this study were developed from multiple sources such as the researchers, experts and literature. Compared with the mere use of inductive and deductive method, the mixed method with multiple item sources will have higher level of validity. Based on the 19 subcategories defined by the grounded theory analysis and some well accepted tourism perceived value scales, a total of 25 initial scale items were developed. Following Delphi method, experts from Guilin Culture and Tourism Bureau, Guangxi Ethnic Tourism Museum and Guilin Tourism University were invited to review and purify the initials items on the principle of conciseness, completeness and correctness. As a result, six items were finally deleted, and some remaining items were refined in wording. The numbers of the items were consistent with those in the formal questionnaire.

4 The Process and Result of the Pilot Survey

4.1 Investigation Process and Sample Profile

The pilot survey questionnaire is divided into three parts. The first part includes a brief description of the questionnaire and the concept of local knowledge tourism with examples and photos. The second part is the 5-dimension measurement adopting the 5-level Likert scale with "1" meaning "strongly disagree" and "5" meaning "strongly agree". The third part addresses the personal information of the correspondent. According to Klein et al. [49] that the sample size should be 10 to 20 times of the number of measured items [49].

The pilot study was carried out through the internet between April and May of 2022. In the midst of the Covid-19 outbreak, an online survey has the potential to significantly cut down on the costs of questionnaire collecting as well as the hazards associated with person-to-person interaction. Secondly, compared with the face-to-face questionnaire distribution and filling, the online respondents may better fill in the form anonymously and show their real attitude. One key recruitment requirement for the respondents is that he or she must have a local knowledge tourism experience as defined by this study in Guilin within the last 12 months. To ensure the reliability and validity, questionnaires that took less than 120 s to complete and all choice are the same are deemed to be invalid. As such 183, or 87.6% of the valid questionnaires were finally collected out of the total 209 distributed. It is generally accepted when the samples number has reached 5–10 times of the items number, it will suffice for

Table 3 Example of open coding of local knowledge tourism perceived value

Concepts	Source	Sample original text
A1 daily application	Interview participants; Travelogue	1. Local people know how to adapt to local conditions. This kind of traditional stilted house[a] design can save land and be built along the slope up. It's warm in winter and cool in summer, and the ground floor is used to keep livestock down here. (T1[b]) 2. Look at the aunt's hair! She is 68 years old but has not a single white hair on her head. In the Yao Long Hair Village, women cut their hair only three times in their whole life, at birth, marriage and death. The aunt said that their habit is to wash their hair with a sort of special cold water that has been used for rice washing all the year round, so their hair is always so black and shiny. (The MFW 2) 3. The secret of the beautiful black hair of the Red Yao lady is that every family in the village has a huge pot for holding the water of rice washing. Put in some Chinese herbal medicines, after baking, heating, and fermentation, the special Yao "shampoo" is finally made. They (The local Yao women) use the unique shampoo for the whole life and so have no white hair and delay the aging process. The secret has been passed on from generation to generation. (MFW 3)
A2 Physical and mental pleasure	Interview participants; Travelogue	1. The Zhuang ethnic people in Longji is a representative of the Northern Zhuang, with unique costumes and customs. Here, we can see the ancient Zhuang folk dances and costumes that has been very well protected. We can also enjoy the beautiful Zhuang folk songs and experience their life scenes such as how they repair their wooden houses. It is so pleasant and refreshing. (T6) 2. We went to various scenic spots on the second day. Everybody was so happy as we could feel and experience the ethnic and local characteristics. (T8) 3. My favorite is the small cradle chair on the balcony. On Saturday night when I got tired and back to the room, I sat on the chair enjoying the authentic ethnic dancing and songs downstairs, the feeling was so nice. (QNE 1)
A3 Perceive the Cost	Interview participants	1. At the beginning we took a bus, because the local terrain was special and tough. We had to get off and walk in the middle of the journey, our bus could not get to the village. We need change to a small local car to reach the final destination at the end. (T6) 2. Longji terraced fields are beautiful but tiring to climb. (T2) 3. It's not easy to get around here, quite challenging and hard. (T3)

(continued)

Table 3 (continued)

Concepts	Source	Sample original text
A4 Feel the atmosphere	Interview participants; Travelogue; News report	1. The children in the kindergarten are now required to wear the ethnic costume during Zhuang March 3rd Festival, giving us a good sense of ritual and a feeling of the festival. (T7) 2. Throwing the embroidered ball is a traditional sport popular with the Zhuang people, and it is also conducted as a ritual to seek a husband. The plot of show can really motivate audience's emotions. (MFW 8) 3. In the live broadcast at the East and West Alley, the hostess showed the special local cuisine, traditional culture and knowledge of Guilin to the audience online, all participants, regardless of their origin and location, can feel the strong festive atmosphere. (Sohu 2)
A5 Inspire association	Interview participants; Travelogue	1. It's amazing!!! Imagine how hard it was for the ethnic locals in the past to dig up so many terraced fields. It's a crazy job that could only be achieved by efforts of generations. (T9) 2. The way they farm reminds me of my childhood life in the countryside. We all grew rice, but some methods and technologies were different from here in the terrace field. (T12) 3. When entering the Yangshuo ancient village, the first thing you see is a variety of shops for local wine, umbrella, pickles, mask, and rouge etc. The manual workshops and old stories all pulled us back to the childhood memories. Those old days at home flashed back in my mind. (Sohu 1)
A6 Perception and thrill	Interview participants; Travelogue	1. Really want to know how they turned this place into farmland, the terraces are so well protected and in use for thousands of years. It's really amazing. (T20) 2. The traditional Miao costumes really go well with the landscape in the terraced fields. Look down from here, it is marvelous to see the great job of their ancestor. (T15) 3. The technologies of the ancient Canal Lingqu and the legendary story of Liu Sanjie are all highlight of the Guilin tour. It's so amazing and a real shock to your heart and soul. (Sohu 3)

(continued)

Table 3 (continued)

Concepts	Source	Sample original text
A7 Identity and attachment	Interview participants; Travelogue	1. I truly admire the extraordinary wisdom of the ancient, how could they manage to overcome the difficulty and achieve the harmony between nature and man. (T8) 2. The image of Fairy of Songs is deeply rooted in people's hearts. Recorded and imagined by people with different social identities background. The huge statue of songs fairy is integrated with the landscape of Guilin, and it is the embodiment of the truth, goodness and beauty of the land. (Sohu 5) 3. It has been ten years since I left Guilin. When I visited my hometown again, I ate the traditional horseshoe cake and saw how people made it, which remains the same as before. Hometown is still in my heart. (LRB 2)
A8 Make friends	Interview participants; Travelogue	1. Such tourism activities are sort of like having a class at school, where people interact and make friends together. (T8) 2. At night I sat on a stool outside the house and talked with strangers from all over the country. I got to know a nice guy from Dongguan who would go to Longji Terrace several times a year for photography. He opened a bottle of wine, and we chatted while drinking, added friends on Wechat, and now we have become good friends. (T15)
……	……	……
A18 Realize personal aspiration	Interview participants	1. I have always been interested in ethnic culture and knowledge. Today I came here to experience these ethnic cultures with my own eyes, which is more vivid and direct than from books. I guess I got what I wanted. (T16) 2. I have seen something very special here, I feel that my life has become a little richer. (T12)
A19 Individual potential	Interview participants; Travelogue	1. Only when we reached the top of the mountain can we truly feel the ritual sense of measuring the land with our own footsteps and the sense of conquest when we look back. (T12) 2. (The local culture study tour in the ancient Dongli Village of Guilin) It was the first time for Xiao Xiao to travel without her family, I always believe that the potential of children is unlimited, we should believe and support them, it's really an unforgettable journey of research and learning. (MFW 6)

[a]The stilt house is a traditional residence house for ethnic minority such as Zhuang, Yao and Miao in southwest of China

[b]Coding rule: T 10 for Tourist No. 10. Secondary data were extracted from online travel agency such as Mafengwo, Qunaer, Little Red Book and news portal such as Sohu, so MFW 3 refers to travelogue No. 3 from Mafengwo. Likewise QNE stands for Qunaer, LRB for Little Red Book etc

Table 4 Results of axial coding and selective coding

Main category	Corresponding subcategory	Value connotation
Functional value (FV)	Daily usage	Perceive the basic utility, functionality and practicability of local knowledge in daily life and the cost to attain such tourism experience
	Physical and mental pleasure	
	Perceive the cost	
Emotional value (EV)	Feel the atmosphere	Obtain the emotions such as happiness, pleasure, attachment, love, admiration, shock, nostalgia, comfort concerned with local knowledge in the tourist experience
	Inspire association	
	Perception and thrill	
	Identity and attachment	
Social value (SV)	Make friends	Local knowledge tourism experience helps one make more friends, deepen the host–tourist interaction and attain other social purpose
	Host–guest interaction	
	Deepen the emotion	
	Obtain the information	
Cognition value (CV)	Experience the culture	Acquire knowledge about the unique ecology, nature, humanity, history, art, religion and other aspects of the tourist destination as a result of the local knowledge tourism experience
	Understand the nature	
	Understand the society	
	Self-cognition	
Self-accomplishment value (SAV)	Self-promotion	Realize personal dream, aspirations, wishes and interests, and raise one's social prestige, knowledge and spirit because of the local knowledge tourism experience
	Social prestige	
	Realize personal aspiration	
	Individual potential	

exploratory factor analysis. As the pilot survey included 5 dimensional variables and 19 items, the sample number therefore has met the requirement.

As shown in Table 5, 56.8% of the respondents are female. 82.5% are 21–60 years old, while those below 20 and above 61 accounted for 5.5% and 2.2% respectively. As for occupations, civil servants represent 22.4%, followed by staff members in cultural, education, science and technology sectors (18.0%), enterprise employees (13.1%) and students (17.5%). Regarding education background, 50.8% are college/undergraduate students. To sum up, the demographic characteristics of the pilot survey respondents are basically in line with the real situation of local knowledge tourists.

4.2 Descriptive Statistical Analysis

According to the descriptive statistical analysis of 183 questionnaires in the pilot survey, the mean values of functional value, emotional value, social value, cognitive

Table 5 Profile of pilot survey respondents (N = 183)

Variable	Category	Frequency	Percentage (%)
Gender	M	79	43.2
	F	104	56.8
Age	Below 20	10	5.5
	21–30	53	29.0
	31–40	43	23.5
	41–50	51	27.9
	51–60	22	12.0
	Above 61	4	2.2
Occupation	Civil servant	41	22.4
	Company staff	24	13.1
	Student	32	17.5
	Teacher/researcher/medical worker	33	18.0
	Others	53	29.0
Education	Junior high school and below	1	0.5
	Senior high school	8	4.4
	Higher diploma/bachelor	93	50.8
	Master and above	80	43.7
	Others	1	0.5
Monthly income (Yuan). Note: USD 1 ≈ Yuan 7.1	Below 2000	30	16.4
	2000–4999	36	19.7
	5000–7999	50	27.3
	8000–9999	22	12.0
	Above 10,000	45	24.6

value and self-actualization value were 4.30, 4.04, 4.15, 4.19 and 3.16, respectively. While the score of self-actualization value was relatively low as compared to other value dimensions, which can be explained by the classic theory of Maslow's needs hierarchy. Self-actualization lies at the top of the hierarchy of needs; hence it is the most complex and difficult to attain. Secondly, the kurtosis and skewness of data are used for testing whether samples are normally distributed, which indicates the universality of samples. As illustrated in Table 6, the data skewness coefficient of each measurement item ranges between −1.903 and 0.156 with the absolute value smaller than 2, and the kurtosis coefficient is between −0.643 and 5.443 with absolute value less than 8, which meet the criteria of normal distribution (Table 7).

Table 6 Descriptive statistical analysis of pilot survey (N = 183)

	Item	Single mean	Overall mean	Standard deviation	Kurtosis	Skewness
Functional value (FV)	FV1	4.08	4.30	0.931	−1.073	1.210
	FV2	4.50		0.718	−1.903	5.443
	FV3	4.30		0.713	−1.064	2.158
Emotional value (EV)	EV1	4.05	4.04	0.768	−0.388	−0.427
	EV2	4.26		0.746	−0.789	0.266
	EV3	4.08		0.784	−0.492	−0.303
	EV4	3.77		0.848	−0.295	−0.474
Social value (SV)	SV1	4.14	4.15	0.742	−0.479	−0.286
	SV2	4.02		0.763	−0.262	−0.643
	SV3	4.34		0.707	−0.968	1.003
	SV4	4.11		0.726	−0.169	−1.079
Cognition value (CV)	CV1	4.26	4.19	0.693	−0.602	0.025
	CV2	4.23		0.722	−0.565	−0.273
	CV3	4.26		0.667	−0.458	−0.256
	CV4	3.99		0.832	−0.616	0.543
Self-accomplishment value (SAV)	SAV1	3.50	3.16	0.905	−0.124	−0.455
	SAV2	2.98		1.107	0.156	−0.427
	SAV3	2.99		1.066	0.121	−0.348
	SAV4	3.18		1.003	−0.006	−0.233

4.3 Reliability Test and Exploratory Factor Analysis

The reliability of the questionnaire was tested by internal consistency coefficient and item-overall correlation coefficient (CITC). The Cronbach's α coefficient and CITC value of each variable were obtained by using SPSS24.0 for the items of five latent variables: functional value, emotional value, social value, cognitive value and self-actualization value. The results showed that the Cronbach's α coefficient of the total scale was 0.914, and of the five subscales were 0.767, 0.784, 0.826, 0.871 and 0.927 respectively, which are all acceptable. The CITC value of each item in the study also reaches the standard of 0.5. However, the Cronbach's α coefficient was significantly increased after the fourth item of cognitive value, i.e. "*I can better reflect on and recognize my own culture through local knowledge tourism experience*", was deleted. To ensure the internal consistency of the questionnaire, it is therefore decided to delete the item. The author argues that the reason why the deleted 4th item of the cognitive value dimension failed to meet the standard is because the respondents in the interview were all Chinese, they may not be able to distinguish between their own culture and that of the others in the case of Guilin.

Table 7 Reliability analysis of the pilot survey questionnaire (N = 183)

Variable	Item	CITC	Cronbach's α after deleting CV 4	Subscale Cronbach's α
Functional value (FV)	FV1	0.539	0.754	0.767
	FV2	0.684	0.567	
	FV3	0.560	0.698	
Emotional value (EV)	EV1	0.576	0.734	0.784
	EV2	0.584	0.730	
	EV3	0.693	0.672	
	EV4	0.507	0.773	
Social value (SV)	SV1	0.590	0.807	0.826
	SV2	0.696	0.758	
	SV3	0.668	0.772	
	SV4	0.648	0.781	
Cognition value (CV)	CV1	0.741	0.814	0.871
	CV2	0.784	0.794	
	CV3	0.789	0.797	
	CV4	0.567	0.894	
Self-accomplishment value (SAV)	SAV1	0.769	0.923	0.927
	SAV2	0.817	0.909	
	SAV3	0.884	0.885	
	SAV4	0.852	0.897	

To investigate the underlying dimensions of the items, an exploratory factor analysis (EFA) with a component analysis and varimax rotation method was carried out. The Kaiser–Meyer–Olkin test (KMO = 0.866) and Bartlett's test of sphericity ($\chi^2 = 5860.12$, df = 528; $p < 0.001$) indicated suitability to conduct the factor analysis. It was found that the remaining 18 items could be extracted into 5 dimensions, and the cumulative explained variance has reached 73.374%. The distribution of loading values of each factor after rotation was reported in Table 8 below.

Based on the above reliability and validity tests, after item CV4 has been deleted, the final 18-item scale is composed of 5 dimensions. All factor loading is more than 0.6 with most loading exceeding 0.7 and hence indicates that the samples have satisfactory convergence validity. The exploratory factor analysis results are basically consistent with the findings of the grounded theory coding, which preliminarily verifies that this newly developed local knowledge tourism perceived value scale has good reliability and validity.

Table 8 Exploratory factor analysis of the pilot survey questionnaire (N = 183)

Variable	Factor structure					
	No	SAV	SV	CV	EV	FV
Functional value (FV)	FV1	0.299	0.094	0.092	0.053	0.735
	FV2	0.076	0.156	0.281	0.196	0.782
	FV3	−0.016	0.071	0.091	0.161	0.815
Emotional value (EV)	EV1	0.127	0.180	0.252	0.639	0.244
	EV2	−0.092	0.215	0.199	0.715	0.225
	EV3	0.137	0.163	0.283	0.776	0.119
	EV4	0.359	0.201	0.030	0.684	−0.034
Social value (SV)	SV1	0.101	0.629	0.272	0.229	0.219
	SV2	0.282	0.773	0.243	0.151	−0.036
	SV3	0.083	0.814	0.112	0.185	0.155
	SV4	0.307	0.703	0.124	0.198	0.088
Cognition value (CV)	CV1	0.103	0.214	0.807	0.284	0.127
	CV2	0.155	0.189	0.836	0.148	0.244
	CV3	0.160	0.234	0.824	0.247	0.131
Self-accomplishment value (SAV)	SAV1	0.811	0.247	0.100	0.155	0.086
	SAV2	0.874	0.143	0.062	0.075	0.101
	SAV3	0.896	0.120	0.117	0.124	0.103
	SAV4	0.889	0.152	0.156	0.061	0.082

5 The Process and Results of Formal Survey

5.1 Investigation Process and Sample Characteristics Analysis

The sample size was finally set at 400 by considering important factors such as the size of Guilin City, the number of measured items, and the time period when the survey is conducted. The survey was carried out in a mixed way. The face-to face survey was done at five major scenic areas of Guilin, including the East and West Alley, Longsheng Rice Terraces, Guangxi Ethnic Tourism Museum etc. To ensure the reliability and validity, the questionnaire was distributed by the authors and 3 undergraduate students of tourism management to facilitate the respondents. And the internet survey was conducted with the online survey firm Wenjuanxin from June to July 2022. A total of 456 questionnaires were collected, including 256 from the site and 200 from online. After excluding the questionnaires with short filling time and consistent answers for all questions, 412 valid questionnaires were finally obtained, with an effective rate of 90.35%. The sample size has fully reached the standard and can be used for subsequent confirmatory data analysis following the suggestions by Hair et al. (2009) [50]. Only those who have a local knowledge tourism experience in Guilin over the last 12 months were invited for the survey. 57.3% of the respondents were male and 93.7% are aged between 21 and 50, which indicates that young and middle-aged tourists show more interest in local knowledge tourism. Local knowledge tourism seems to be more popular among tourists with higher education background as 70.9% of the respondents had a college/university degree, this may be related to its knowledge-oriented attributes. Due to the ongoing Covid-19 pandemic, 57% of the respondents were from Guangxi and the rest mainly from the neighbouring provinces.

5.2 *Descriptive Statistical Analysis of the Formal Survey*

The descriptive statistical analysis of 412 questionnaires in the formal survey suggests that the overall mean values of functional value, emotional value, social value, cognitive value and self-actualization value were 3.98, 3.80, 3.79, 3.92 and 3.45, respectively. The score of self-actualization value was still relatively low, but higher that 3.16 of the pilot surveys.

5.3 *Normal Distribution Test and Common Variance Test*

As shown in Table 9 the data skewness coefficient of each measurement item is between –0.931 and –0.022 with absolute value less than 2, and the kurtosis coefficient is between –0.943 and 1.481 with absolute value less than 8, which indicates that samples are normally distributed. The factor analysis suggested that the explained variance of the first factor without rotation was 27.410%, smaller than 40%, and there was no large amount of explained variance concentrated in one factor, indicating that the influence of common method deviation was small and acceptable.

5.4 *Analysis of Model Fitting*

This study constructed a five-factor model of tourist perceived value as Fig. 1 and used AMOS25.0 to conduct the confirmatory factor analysis. As illustrated in Table 10. $\chi 2/df$ is 1.505, which is between 1 and 3, and significant at the level of $p < 0.001$. RMR is 0.033, less than 0.05; RMSEA is 0.035, less than 0.08. Other fitting indices such as GFI, AGFI, NFI, RFI, IFI, TLI, CFI have all reached the corresponding standards, indicating that it is an acceptable model.

Table 9 Descriptive statistical analysis of the formal survey

	Item no	Item mean	Dimension mean	Standard deviation	Kurtosis	Skewness
Functional value (FV)	FV1	3.94	3.98	0.798	−0.931	1.481
	FV2	4.05		0.889	−0.763	0.310
	FV3	3.96		0.855	−0.813	0.861
Emotional value (EV)	EV1	3.90	3.80	0.829	−0.421	−0.328
	EV2	3.75		0.826	−0.317	−0.246
	EV3	3.81		0.819	−0.353	−0.322
	EV4	3.75		0.832	−0.255	−0.347
Social value (SV)	SV1	3.90	3.79	0.850	−0.167	−0.909
	SV2	3.69		0.770	−0.026	−0.462
	SV3	3.83		0.863	−0.076	−0.943
	SV4	3.74		0.794	−0.116	−0.344
Cognition value (CV)	AV1	3.93	3.92	0.840	−0.258	−0.755
	AV2	3.87		0.878	−0.283	−0.738
	AV3	3.97		0.847	−0.432	−0.514
Self-accomplishment value (SAV)	MV1	3.80	3.45	0.858	−0.694	0.559
	MV2	3.12		1.077	−0.022	−0.611
	MV3	3.39		1.107	−0.385	−0.510
	MV4	3.50		1.057	−0.466	−0.406

5.5 *Reliability and Validity Tests*

Cronbach's α coefficient and combined reliability CR value are used to test the reliability of each latent variable in the theoretical model to determine the consistency and stability of the data. As shown in Table 11, the Cronbach's α coefficient of the total scale has reached 0.840, and ranges from 0.810 to 0.850 for each latent variable (>0.7). At the same time, the CR value of the combined reliability of each latent variable was between 0.803 and 0.850 (>0.7). According to the reliability test standard proposed by Fornell and Larcker (1981), the measurement of each latent variable has good reliability (Table 12).

The validity test is conducted from two aspects: aggregate validity and discriminant validity. As can be seen from Table 13, the standardized factor loading of each measurement item is between 0.643 and 0.866 (>0.5) and AVE values of each measurement item ranges from 0.505 to 0.655 (>0.5), which indicates the good aggregate validity. Regarding the discriminant validity, the square root of the extracted average variance of each latent variable is higher than the correlation coefficient between each latent variable (see Table 13 for details), suggesting that the discriminant validity is also satisfactory.

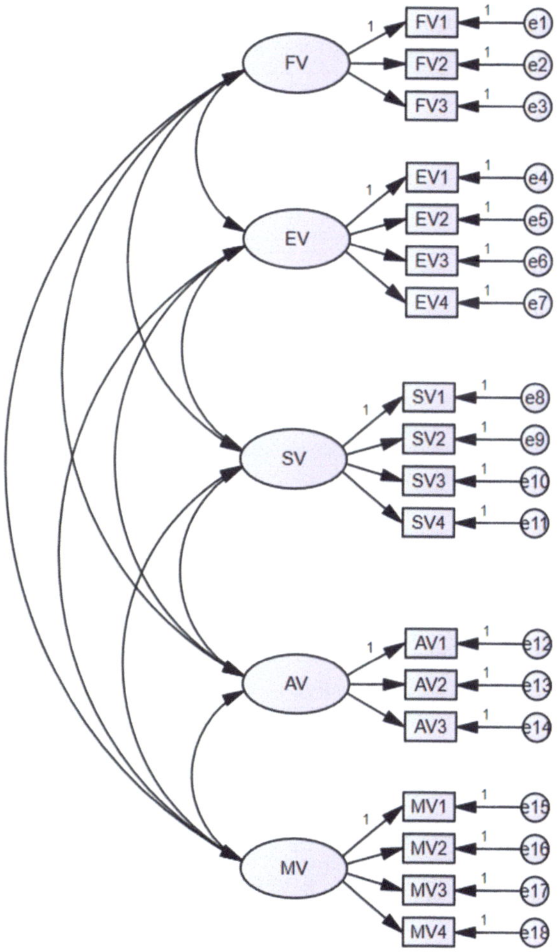

Fig. 1 Five-factor model of tourist perceived value

Through the above steps of confirmatory factor analysis, the items of the local knowledge tourism experience value measurement scale are finally confirmed as in Table 13.

Table 10 Fitting results of the measurement model (N = 412)

Model fitting index	Testing data	Judgment threshold
χ^2/df	1.505	$1 < \chi^2/df < 3$
RMR	0.033	RMR < 0.05
RMSEA	0.035	RMSEA < 0.08
GFI	0.951	0.9 < GFI < 1
AGFI	0.934	0.8 < AGFI < 1
NFI	0.936	NFI > 0.9
RFI	0.922	RFI > 0.9
PNFI	0.765	PNFI > 0.5
PGFI	0.696	PGFI > 0.5
IFI	0.978	IFI > 0.9
TLI	0.972	TLI > 0.9
CFI	0.977	CFI > 0.9

Table 11 Reliability test of the measurement model (N = 412)

	Item No	Standardized factor loading	AVE	CR	Cronbach's α
Functional value (FV)	FV1	0.866	0.655	0.850	0.847
	FV2	0.783			
	FV3	0.776			
Emotional value (EV)	EV1	0.710	0.516	0.810	0.809
	EV2	0.703			
	EV3	0.755			
	EV4	0.704			
Social value (SV)	SV1	0.778	0.505	0.803	0.801
	SV2	0.666			
	SV3	0.748			
	SV4	0.643			
Cognition value (CV)	AV1	0.798	0.607	0.822	0.820
	AV2	0.723			
	AV3	0.813			
Self-accomplishment value (SAV)	MV1	0.664	0.548	0.828	0.825
	MV2	0.740			
	MV3	0.746			
	MV4	0.803			
The total scale					**0.840**

Table 12 Discriminant validity test (N = 412)

	FV	EV	SV	MV
FV	**0.809**			
EV	0.410***	**0.718**		
SV	0.345***	0.289***	**0.711**	
AV	0.285***	0.378***	0.285***	
MV	0.212***	0.299***	0.279***	**0.740**

6 Discussion and Conclusion

6.1 Structure and Features of Local Knowledge Tourism Perceived Value

Based on the theory of customer perceived value and the case study of Guilin, Guangxi China, this study has first identified 5 perceived value dimensions that include functional value, emotional value, social value, cognitive value and self-actualization value of local knowledge tourism using the grounded theory. Later the rigorous procedures have been followed to develop the related scale that consists of 18 items as illustrated in Table 13. Related tests indicate that the newly develop scale is of satisfactory reliability and validity. This study attempts to probe into the tourist perceived value of local knowledge to fill in the gap, which is conducive to its conservation and development. The development of local knowledge as a tourist attraction can highlight the distinct features and local spirit of the destination, thus helping to strengthen the destination competitiveness.

Inspired by hierarchy of needs, Pearce and Caltabiano (1983) studied the progressive model of tourist demand and found that the higher the level of tourism motivation, the higher the demand will be [51]. The local knowledge tourism perceived values identified by grounded theory in this study also showed corresponding progressive relationship between the tourist perceived values dimensions. As shown in Fig. 2, the five identified local knowledge tourism perceived value dimensions form a corresponding internal relationship in line with the Maslow's hierarchy of needs. The functional value is reflected in the tourist perception that local knowledge can be utilized to meet the basic physiological needs such as food, accommodation and transportation. Emotional value related to a variety of positive emotions that have been triggered in the tour. Local knowledge experience can stimulate tourists' curiosity, association and nostalgia, which may be transferred into emotions and feelings such as place identification and attachment to the destination and enhance tourists' sense of belonging and safety. Social value refers to the social benefits such as making more new friends through the study of local knowledge in tourism activities. Other benefits may include the deepening of host–tourist interaction, closer ties between friends and improved family relationship, which is corresponding to the satisfaction of the belonging and love needs in the Maslow's hierarchy. The major motivation

Table 13 Items of the local knowledge tourism perceived value scale

Construct	Item	Reference and source
Functional value (FV)	1. I can perceive the usage and function of local knowledge in daily life such as cuisine, living, traveling and medical treatment in the tour	Na et al. [46], interview and observation
	2. The local knowledge tourism experience makes me feel physically and mentally happy	
	3. This tour experience entails special financial, time and energy costs	
Emotional value (EV)	4. I can feel the strong atmosphere created by the traditional culture, festivals, history and legends of the tourist destination	Sheth et al. [38], interview and observation
	5. I may associate the past era and the scene of people's life at that time due to the local knowledge tourism experience	
	6. I am amazed at the miracle of local knowledge and traditional wisdom in the destination	
	7. I may have the feeling of place identity and attachment with the destination because of the local knowledge tourism experience	
Social value (SV)	8. Local knowledge tourism can help me make more friends	Ma and Bao [45], interview and observation
	9. The local knowledge tourism experience makes me more willing to communicate and interact with the local people	
	10. Participating in local knowledge tours with friends and family can enhance mutual understanding and affection between us	
	11. Local knowledge tourism experience provides me with more information and knowledge about the destination	
Cognition value (CV)	12. I can better understand the unique history, tradition, culture and art of the destination from the local knowledge tourism experience	Interview and observation
	13. I can better understand the unique ecological and natural environment of the destination through local knowledge tourism experience	
	14. I can better understand the unique social and humanistic environment of the destination through local knowledge tourism experience	

(continued)

Table 13 (continued)

Construct	Item	Reference and source
Self-accomplishment value (SAV)	15. Local knowledge tourism experience makes me a better self	Interviews and observations
	16. Local knowledge tourism experience can enhance my social prestige	
	17. Local knowledge tourism experience enables me to realize my aspiration and ambition	
	18. Local knowledge tourism experience allows me to exert my potential	

of local knowledge tourism is to seek and acquire knowledge. As such, the tourism activities will enhance participants' understanding of the destination with regards to its natural environment and traditional culture and promote the mutual understanding and respect between hosts and guests. Finally, the self-actualization value implies that some tourists manage to enhance their cultural literacy and realize their personal dreams and aspirations in the local knowledge tourism experience. This corresponds to the highest level of self-actualization in the Maslow's hierarchy. As a tourism activity typically focusing on cultural experience, the perceived value of local knowledge tourism presents a systematization and multi-dimensional features. With the deepening of tourism experience and host-tourist interaction, the tourist perceived value will evolve step by step from the satisfaction of basic philological needs to the advanced level such as self-actualization to meet the emotional, social, and self-actualization needs of the tourist.

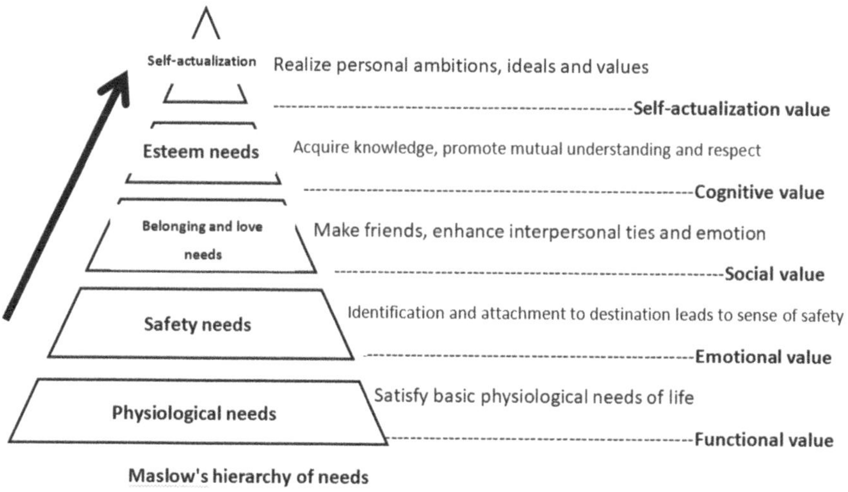

Fig. 2 Hierarchical model of local knowledge tourism perceived values

6.2 Research Implication

As an important concept in the field of consumer experience quality, perceived value has been extensively used in tourism research. Identifying and improving tourist perceived value is an important antecedent for upgrading the competitiveness of tourist destinations. In theory, this study puts forward a new concept of local knowledge tourism from the category of heritage tourism and addressed the perceived value in this setting, which expands the research scope of heritage tourism and enrich the research context of tourism perceived value. At the same time the newly developed scale will add to the understanding of the connotation, components and distinct features of local knowledge tourism in the lens of tourism experience and perceived value. The scale can be a useful tool for subsequent study such as the impacts of local knowledge perceived values on tourist behaviors.

In management practice, the study will underline the unique values of local knowledge in tourism development, enhance cultural tourism product development and the comprehensive competitiveness of the destination. The tourist perceived values identified will reveal the various dimensions and levels of local knowledge tourism experience, hence providing insights on market segmentation and marketing strategy formulation for this niche cultural tourism market. For example, according to the research, local knowledge tourists can be divided into functional experience type, emotional attachment type, knowledge seeking and development type and self-actualization type. From the perspective of tourists, this study is conducive to improve their tourist experience, strengthen their participation, and deepen their understanding of the tradition and culture of the destination.

6.3 Limitations and Future Research

Despite a number of contributions, this study still has some limitations. Firstly, local knowledge varies a lot in forms and types, and so do the tourist when it comes to their background, motivation, preference and demand. To broaden this exploratory study, future research can be further conducted under different local knowledge tourism settings. The comparative study can be expanded to tourists of different age, occupation, education and income level. Secondly, this research is primarily conducted after the outbreak of the pandemic, China's national border remains close for international tourists up to the end of the study. As a consequence, the interview participants and online travelogue used as secondary data were all Chinese. When China's inbound markets is restored, the in-depth interviews can be conducted with international tourists of different cultural and social background.

References

1. Lenzerini, F.: Intangible cultural heritage: the living culture of peoples. Eur. J. Int. Law **22**(1), 101–120 (2011). https://doi.org/10.1093/ejil/chr006
2. Hill, R., Adem, Ç., Alangui, W.V., Molnár, Z., Aumeeruddy-Thomas, Y., Bridgewater, P., Tengö, M., Thaman, R., Adou Yao, C.Y., Berkes, F., Carino, J., Carneiro da Cunha, M., Diaw, M.C., Díaz, S., Figueroa, V.E., Fisher, J., Hardison, P., Ichikawa, K., Kariuki, P., Xue, D.: Working with Indigenous, local and scientific knowledge in assessments of nature and nature's linkages with people. Curr. Opin. Environ. Sustain. **43**, 8–20 (2020). https://doi.org/10.1016/j.cosust.2019.12.006
3. Holifield, R.: Environmental justice as recognition and participation in risk assessment: negotiating and translating health risk at a superfund site in Indian country. Ann. Assoc. Am. Geogr. **102**(3), 591–613 (2012). https://doi.org/10.1080/00045608.2011.641892
4. Ramos, H.: Mapping the field of environmental justice: redistribution, recognition and representation in ENGO press advocacy. *The Canadian Journal of Sociology/Cahiers Canadiens de Sociologie*, 40(3), 355–376 (2015). JSTOR.
5. Overstreet, G.A.: Creating a value in oversupplied markets: the case of Charlotesville, Vinia, Hotels. Cornell Hotel and Restaurant Administration Quarterly **34**(5), 68–96 (1993). https://doi.org/10.1177/001088049303400513
6. Li, J., Cheng, S., Zhong, L.: Progress on the research of customer value theory in tourism. Hum. Geogr. **24**(5), 21–24 (2009)
7. Geertz, C.: From the native's point of view": on the nature of anthropological understanding. Bull. Am. Acad. Arts Sci. 26–45 (1974).
8. UNESCO: Local Knowledge, Global Goals. UNESCO. https://en.unesco.org/lik-expo (2017).
9. Lepore, W., Hall, B.L., Tandon, R.: The knowledge for change Consortium: a decolonising approach to international collaboration in capacity-building in community-based participatory research. Canadian Journal of Development Studies/Revue Canadienne d'études Du Développement **42**(3), 347–370 (2021)
10. Trigger, B.G.: Alternative archaeologies: nationalist, colonialist. Imperialist. Man **19**(3), 355–370 (1984). https://doi.org/10.2307/2802176
11. Gao, J., Wu, B.: Revitalizing traditional villages through rural tourism: a case study of Yuanjia Village, Shaanxi Province, China. Tour. Manage. **63**, 223–233 (2017)
12. Kneafsey, M.: Rural cultural economy: tourism and social relations. Ann. Tour. Res. **28**(3), 762–783 (2001)
13. Quinn, B.: Arts festivals and the city. Urban Studies **42**(5–6), 927–943 (2005)
14. Glover, N.: Co-produced histories: Mapping the uses and narratives of history in the tourist age. Public Hist. **30**(1), 105–124 (2008)
15. Yang, L., Wall, G., Smith, S.L.: Ethnic tourism development: Chinese Government Perspectives. Ann. Tour. Res. **35**(3), 751–771 (2008)
16. de Bruin, A., Jelinčić, D.A.: Toward extending creative tourism: participatory experience tourism. Tourism Rev. **71**(1), 57–66 (2016). https://doi.org/10.1108/TR-05-2015-0018
17. Pongponrat, K.: Participatory management process in local tourism development: a case study on fisherman village on Samui Island, Thailand. Asia Pacific J Tourism Res **16**(1), 57–73 (2011). https://doi.org/10.1080/10941665.2011.539391
18. Pongponrat, K., Chantradoan, N. J.: Mechanism of social capital in community tourism participatory planning in Samui Island, Thailand. *Tourismos*, 7(1) (2012)
19. Chen, H., Rahman, I.: Cultural tourism: An analysis of engagement, cultural contact, memorable tourism experience and destination loyalty. Tourism Manage. Perspect. **26**, 153–163 (2018). https://doi.org/10.1016/j.tmp.2017.10.006
20. Butler, R., & Hinch, T.: Tourism and indigenous peoples: issues and implications. Routledge (2007)
21. Beritelli, P., Bieger, T., Laesser, C.: Destination governance: using corporate governance theories as a foundation for effective destination management. J. Travel Res. **46**(1), 96–107 (2007)

22. Halkier, H.: Innovation and destination governance in Denmark: tourism, policy networks and spatial development. Eur. Plan. Stud. **22**(8), 1659–1670 (2014)
23. Pechlaner, H., Beritelli, P., Pichler, S., Peters, M., & Scott, N. R.: Contemporary destination governance: a case study approach. Emerald Group Publishing (2015).
24. Zeithaml, V.A.: Consumer perceptions of price, quality, and value: a means-end model and synthesis of evidence. J. Mark. **52**(3), 2–22 (1988). https://doi.org/10.1177/002224298805200302
25. Chen, Z., Dubinsky, A.J.: A conceptual model of perceived customer value in e-commerce: a preliminary investigation. Psychol. Mark. **20**(4), 323–347 (2003). https://doi.org/10.1002/mar.10076
26. McDougall, G.H.G., Levesque, T.: Customer satisfaction with services: putting perceived value into the equation. J. Serv. Mark. **14**(5), 392–410 (2000). https://doi.org/10.1108/08876040010340937
27. Sánchez-Fernández, R., Iniesta-Bonillo, M.Ã.: The concept of perceived value: a systematic review of the research. Mark. Theory **7**(4), 427–451 (2007). https://doi.org/10.1177/1470593107083165
28. Snoj, B., Pisnik Korda, A., Mumel, D.: The relationships among perceived quality, perceived risk and perceived product value. J. Prod. Brand Manage. **13**(3), 156–167 (2004). https://doi.org/10.1108/10610420410538050
29. Bao, J., Xie, H.J.: Determinants of domestic tourism demand for Guilin. J. China Tourism Res. **15**(1), 1–14 (2019)
30. Murphy, P., Pritchard, M.P., Smith, B.: The destination product and its impact on traveller perceptions. Tour. Manage. **21**(1), 43–52 (2000). https://doi.org/10.1016/S0261-5177(99)00080-1
31. Stevens, B.F.: Price value perceptions of travelers. J. Travel Res. **31**(2), 44–48 (1992). https://doi.org/10.1177/004728759203100208
32. Chen, C.-F., Chen, F.-S.: Experience quality, perceived value, satisfaction and behavioral intentions for heritage tourists. Tour. Manage. **31**(1), 29–35 (2010). https://doi.org/10.1016/j.tourman.2009.02.008
33. Biao, H., DengMing, X., LianXin, Z., Guo, Q.: Scale development and empirical validation of duty-free shopping tourist perceived value. Tourism Tribune **35**(4), 120–132 (2020)
34. Holbrook, M.B., Batra, R.: Assessing the role of emotions as mediators of consumer responses to advertising. Journal of Consumer Research **14**(3), 404–420 (1987)
35. Woodruff, R.B.: Customer value: the next source for competitive advantage. J. Acad. Mark. Sci. **25**(2), 139 (1997). https://doi.org/10.1007/BF02894350
36. Huang, Y., Huang, F.: Tourists' perceived value model and its measurement: an empirical study. Tourism Tribune **22**(8), 42–47 (2007)
37. Babin, B. J., Darden, W. R., & Griffin, M.: Work and/or fun: Measuring hedonic and utilitarian shopping value. J. Consumer Res. 20(4), 644–656 (1994). https://doi.org/10.1086/209376
38. Sheth, J.N., Newman, B.I., Gross, B.L.: Why we buy what we buy: a theory of consumption values. J. Bus. Res. **22**(2), 159–170 (1991). https://doi.org/10.1016/0148-2963(91)90050-8
39. Sweeney, J.C., Soutar, G.N.: Consumer perceived value: the development of a multiple item scale. J. Retail. **77**(2), 203–220 (2001). https://doi.org/10.1016/S0022-4359(01)00041-0
40. de Ruyter, K., Wetzels, M., Bloemer, J.: On the relationship between perceived service quality, service loyalty and switching costs. Int. J. Serv. Ind. Manag. **9**(5), 436–453 (1998). https://doi.org/10.1108/09564239810238848
41. Grönroos, C.: Value-driven relational marketing: from products to resources and competencies. J. Mark. Manag. **13**(5), 407–419 (1997). https://doi.org/10.1080/0267257X.1997.9964482
42. Groth, J.: Important factors in the sale and pricing of services. Manag. Decis. **33**(7), 29–34 (1995). https://doi.org/10.1108/00251749510090557
43. Petrick, J.F.: The Roles of Quality, Value, and Satisfaction in Predicting Cruise Passengers' Behavioral Intentions. J. Travel Res. **42**(4), 397–407 (2004). https://doi.org/10.1177/0047287504263037

44. Andrades, L., Dimanche, F.: Co-creation of experience value: a tourist behaviour approach. In Creating experience value in tourism (pp. 83–97). CAB International Wallingford UK (2018).
45. Ma, L., Bao, J.: Traditional festival tourism experience from the perspective of perceived value——taking Xishuangbanna Dai water splashing festival as an example. Geogr. Res. **31**(2), 269–278 (2012)
46. Na, M., Xie, Y., & Gursoy, D.: Destination experience value: dimensional identification, scale development and validation. Tourism Tribune 34(12), 48–60 (2019). https://doi.org/10.19765/j.cnki.1002-5006.2019.12.009
47. Pi, P., Liu, X., Gao, Y.: Research on scale development of tourism destination customer experience value. Value Eng. **35**, 30–35 (2016). https://doi.org/10.14018/j.cnki.cn13-1085/n.2016.24.012
48. Mills, J., Bonner, A., Francis, K.: The development of constructivist grounded theory. Int. J. Qual. Methods **5**(1), 25–35 (2006). https://doi.org/10.1177/160940690600500103
49. Klein, K.J., Dansereau, F., Hall, R.J.: Levels issues in theory development, data collection, and analysis. Acad. Manag. Rev. **19**(2), 195–229 (1994). https://doi.org/10.5465/amr.1994.9410210745
50. Hair, J. F.: Multivariate data analysis. Upper Saddle River, NJ: Pearson (2009).
51. Pearce, P.L., Caltabiano, M.L.: Inferring travel motivation from travelers' experiences. J. Travel Res. **22**(2), 16–20 (1983). https://doi.org/10.1177/004728758302200203

Spatio-Temporal Impact of Celebrity Endorsement on Tourist Destination Network Attention: A Case Study

Jinbo Jiang, Huichan Lin, and Yun Long

Abstract This artical takes Ding Zhen endorsing Litang County as an example, and uses Baidu Index to compare and analyze the temporal and spatial evolution characteristics of network attention of destination of celebrity endorsement. There are three major findings. The first one is that there are both temporal synchronization and spatial consistency in network attention of celebrity and network attention of tourist destination. In terms of temporal evolution, immediately after the celebrity endorsement, the network attention to the tourist destination increases significantly. The seasonal difference is reduced and the skewness of distribution within National Day Golden Week tends to be uniform. The last finding is about the spatial characteristics. After celebrity endorsement, the degree of agglomeration of network attention decreases and intra-regional difference and intra-provincial difference also decreases. Besides, this article also provides some suggestions for tourism destinations to effectively use celebrity endorsements.

Keywords Celebrity endorsement · Network attention · Litang County · Destination marketing

1 Introduction

In recent years, the rapid development of the Internet has greatly promoted the transformation of tourism [1]. With the rapid spread of the Internet, more and more tourist destinations have become Internet celebrity check-in places, and many of them have risen rapidly due to celebrity effects [2]. Celebrities attract much attention from the public because of their works, lifestyles as well as personal charm. Celebrity endorsement refers to celebrities using their public influence to represent a product

J. Jiang · H. Lin · Y. Long (✉)
South China University of Technology, Guangzhou, Guangdong, People's Republic of China
e-mail: lyabbyly@163.com

J. Jiang
e-mail: jbjiang@scut.edu.cn

Y. Luo et al. (eds.), *Tourist Experience and Perception in the Pandemic and Post-Pandemic Era*, https://doi.org/10.1007/978-981-96-1820-0_4

[3], and it has become an important way of modern enterprise marketing [4]. After entering the twenty-first century, celebrity endorsements are often used in destination marketing to attract tourists' attention, thereby promoting the development of tourist destinations.

The application of celebrity endorsements in practice has attracted extensive academic attention. Scholars have mainly explored the effectiveness of celebrity endorsements based on source model [5, 6] and matching hypothesis (or consistency model) [7, 8]. The source model studies reliability and attractiveness of celebrity endorsements. The former depends on celebrity's professionalism and trustworthiness [9], while the latter depends on the celebrity's appearance, personality, lifestyle and other characteristics [10]. The source model believes that the effectiveness of celebrity endorsements has nothing to do with the product [11]; the matching hypothesis further considers the brand and consumer factors that are ignored in the source model, and believes that the effectiveness of celebrity endorsements depends on endorser-brand match, endorser-self match or brand-self match [12]. Relevant studies have mainly focused on enterprises and their products, and rarely on tourist destinations [13, 14]. Although the research on celebrity endorsement of tourist destinations is still in its infancy, existing studies have proved the effectiveness of celebrity endorsement in tourist destination marketing, mainly in terms of tourist destination branding and tourists' attitudes and behaviors [12, 15–17]. Scenario experiment method and questionnaire survey method are commonly used. Scenario experiment method, also is defined as quasi-experimental method [16], requires subjects to watch the processed celebrity endorsement posters or other stimulating materials, and then fill in the questionnaire after the experiment [15, 18]. Questionnaire survey method selects areas where celebrity endorsements have been used as the research site [17], and investigates respondents' emotional attitudes and behavioral intention. It can be seen that most of the current research on celebrity endorsements of tourist destinations focuses on the influencing factors and effectiveness of celebrity endorsements from the perspective of tourists, and is conducted through scenario experiments and questionnaire surveys. However, researchers rarely pay attention to temporal and spatial evolution characteristics of destination after celebrity endorsements. Will popularity of tourist destinations be improved after celebrity endorsements? How long does the celebrity effect last? How wide is the scope of influence, and what are the fluctuation trends and spatial distribution differences?

As of June 2021, the scale of Chinese netizens has reached 1.011 billion, with a penetration rate of 71.6%.[1] The Internet has become an important interactive communication medium, which can quickly amplify the effect of celebrity endorsements and promote the rapid development of tourist destinations [19]. The attention of tourist destinations can also be reflected through Internet data, such as network attention. Network attention, which can be replaced by Baidu Index, is a data that reflects the degree of public attention or demand for something by collecting statistics of people's search traces on the Internet [20] under the premise of the popularization

[1] The data comes from the 48th "Statistical Report on Internet Development in China" by China Internet Network Information Center.

of the Internet. Some research has proved that network attention is closely related to passenger flow, and there is a strong positive correlation between the two [21–23]. Hence, in order to elucidate the important effect of celebrity endorsement on tourist destinations, this study examines the case of Ding Zhen endorsing Litang County, utilizing network attention data sourced from Baidu Index. It investigates the temporal and spatial impacts of celebrity endorsements on tourist destinations, and concludes with pertinent recommendations.

The study is developed according to the following structure: Firstly, a brief literature review of celebrity endorsement is reported and research questions are proposed. Secondly, the case background and methodology used for data collection and analysis is explained. The latter is followed by analytical results where temporal variation and spatial distribution characteristics are conveyed. Lastly, conclusions are conducted and suggestions are provided.

2 Case Overview, Data Collection and Research Method

2.1 Case Overview

Litang County is located in the western part of Sichuan Province, with an altitude of more than 4,000 m. It was withdrawn from the poverty-stricken county on February 20, 2020. As early as 2016, Litang County proposed to cultivate tourism as a local leading and pillar industry, and to expand its influence through various methods, such as making tourism promotional videos, creating plateau mountain sports, plateau international horse racing festival and other activities.[2] However, due to its low popularity, high terrain and difficult conditions, the tourism development in Litang County was relatively slow.

On November 11, 2020, a short video about Dingzhen, a Tibetan boy, became popular on the Internet. Many people called him "Sweet Wild Boy" because of his innocent, wild smile and handsome appearance. The launch of video *Ding Zhen's World*, discussion *Where is Litang County where Ding Zhen is located*, and the introduction from spokesperson of the Ministry of Foreign Affairs pushed the hot event to a climax. Topics related to Ding Zhen had been on the Sina Weibo hot search list for many times within a month. At the same time, the Litang County Government hired Ding Zhen as the tourism ambassador of Litang County. According to statistics, in 2020, Litang County received a total of more than 1.512 million tourists and achieved a tourism income of 1.66 billion yuan,[3] which was a big improvement

[2] Litang County People's Government. Litang County has cultivated tourism into a leading and pillar industry. http://www.gzlt.gov.cn/ltxrmzf/c100332/201603/5f00f88c060d4ec0b48ad6e121ffcedc.shtml.

[3] Litang County People's Government. [Reading in one picture] Litang County's 2020 Government Work Report. http://www.gzlt.gov.cn/ltxrmzf/c103247/202012/2d4215e75b1747cab01f97eb1b33b228.shtml, 3–10.

compared to 2019. This study believes that Ding Zhen has high popularity and public influence, which is in line with the definition of celebrity.

2.2 Data Collection

Network attention can be replaced by keyword search volume of netizens on Baidu websites, also known as search index [24]. Combined with the actual inclusion of keywords in Baidu Index website, this article uses the sum of retrieval volume of the words "Ding Zhen" and "Ding Zhen's world" as the network attention of Ding Zhen. Ding Zhen's pony Pearl and Sweet Wild Boy are also celebrity-related words, but are excluded respectively because the meaning of the word "pearl" itself is interfered by the retrieval information and "Sweet Wild Boy" are not include in Baidu Index. The sum of the retrieval volume of the word "Litang" and "Litang County" are regarded as the network attention of the tourist destination of Litang County. The retrieval time of network attention of the celebrity was from November 13, 2020 to December 31, 2021, and in order to better compare the difference before and after the network attention of endorsed tourist destinations, the retrieval time of network attention of the tourist destinations is set from January 1, 2017 to December 31, 2021. The retrieval website is Baidu Index website (http://index.baidu.com/), and the retrieval scope is the whole country and 31 provinces, municipalities and autonomous regions (excluding Hong Kong, Macao and Taiwan regions).

2.3 Analytical Method

2.3.1 Seasonal Concentration Index

Seasonal concentration index is used to measure the monthly change and concentration of network attention. The calculation formula is:

$$R = \sqrt{\sum_{i=1}^{12} \frac{\left(100\frac{X_i}{\sum_i^{12} X_i} - 8.33\right)^2}{12}} \tag{1}$$

In the formula, R is the seasonal concentration index, and X_i is the monthly network attention. The larger the R value, the more concentrated the monthly distribution and the greater the seasonal difference, that is, the more obvious the off-peak season. On the contrary, the closer the R value is to 0, the more uniform the monthly distribution and the smaller the seasonal difference, and that is the less obvious the off-peak season [25].

2.3.2 Intra-Week Distribution Skewness Index

Intra-Week distribution skewness index is used to measure the distribution characteristics of network attention in the golden week. The calculation formula is:

$$T = 100 \times \frac{2}{7}\left(\sum_{i=1}^{7} it_i - 4\right) \tag{2}$$

In the formula, t_i is the percentage of the network attention on the i th day and the network attention in the whole week, and i is the sequence number of the network attention from large to small. $T>0$ indicates that the network attention is concentrated in the late period of the week, $T<0$ indicates that the network attention is concentrated in the early period, and $T=0$ indicates that the network attention is evenly distributed in the week [25].

2.3.3 Gini Coefficient

Gini Index is used to reflect the degree of concentration of the spatial distribution. Its calculation formula is as follows:

$$G = \frac{1}{2m^2 n}\sum\nolimits_{j=1}^{m}\sum\nolimits_{i=1}^{m}\left|X_j - X_i\right| \tag{3}$$

G, m and n in the formula respectively represent the Gini coefficient, number of provinces, municipalities and autonomous regions studied and the average value of the network attention of each province. $\left|X_j - X_i\right|$ represents the absolute value of network attention of any two provinces i and j. The G value is in the range of [0, 1] and the larger the value, the higher the concentration. When the G value is lower than 0.2, the distribution is too equal, and when the G value is higher than 0.4, the distribution is too unequal [26, 27].

2.3.4 Coefficient of Variation

Coefficient of variation is used to measure the degree of regional differences in network attention in Litang County. Its calculation formula is:

$$V_c = \frac{\sqrt{\frac{\sum_{i=1}^{n}(X_i - \overline{X})^2}{n}}}{\overline{X}} \tag{4}$$

In the formula, V_c is the coefficient of variation of each region, n is the number of provinces; X_i is the actual network attention of each province; X_i is the average network attention of each province. The larger the value of V_c, the greater the degree of regional differences and the higher the agglomeration, and vice versa [28].

3 Temporal Variation Characteristics of Network Attention

3.1 Synchronization Analysis of Network Attention of Celebrity and Destination

3.1.1 Long-Term Trends

As shown in Fig. 1, after the keyword "Ding Zhen" was included in Baidu Index website, the network attention of Ding Zhen and Litang County roughly went through three stages: the gestation period, the outbreak period, and the continuation period [20, 29]. November 13 to November 26, 2020 is the gestation period of the two, with the average national network attention 19,775.05 and 2439.07 respectively. November 27 to December 13, 2020 is the outbreak period, with the average network attention 106,779.00 and 5793.41, which increase by 439.97% and 137.53% respectively compared with the gestation period. December 14 to December 31, 2021 is the extension period, and the average national network attention of the two are 7193.86 and 1337.53, which decrease by 93.26% and 76.91% respectively compared with the outbreak period.

3.1.2 Short-Term Trends

Through the comparative analysis of the long-term trend, fluctuation trend of network attention of Dingzhen and Litang County is the most obvious within 100 days. Therefore, network attention of Dingzhen and Litang County within 100 days from November 13 to February 20 is further analyzed (Fig. 2). Within 100 days, the network attention of Ding Zhen and Litang County shows a unimodal trend as a whole, and the trend of the two is basically the same. The peak appears in the outbreak stage, that is, it increases sharply before the peak, and drops sharply after the peak. But small peaks also appears in some periods after the peak. We use IBM SPSS Statistics 22 to conduct a non-parametric correlation test on Ding Zhen's daily network attention and Litang County's daily network attention, and the result shows that the two are significantly correlated (sig = 0.000), indicating that Ding Zhen's popularity on the Internet has a significant impact on Litang County's network attention. Therefore, Ding Zhen's endorsement has improved the popularity of Litang County in a short period of time, and brings unprecedented opportunities for the tourism development of Litang County. In addition, nowadays, network hot events spread faster and information update cycle is shorter [29], so the entire life cycle does not experience an obvious growth period, but converges and bursts in a very

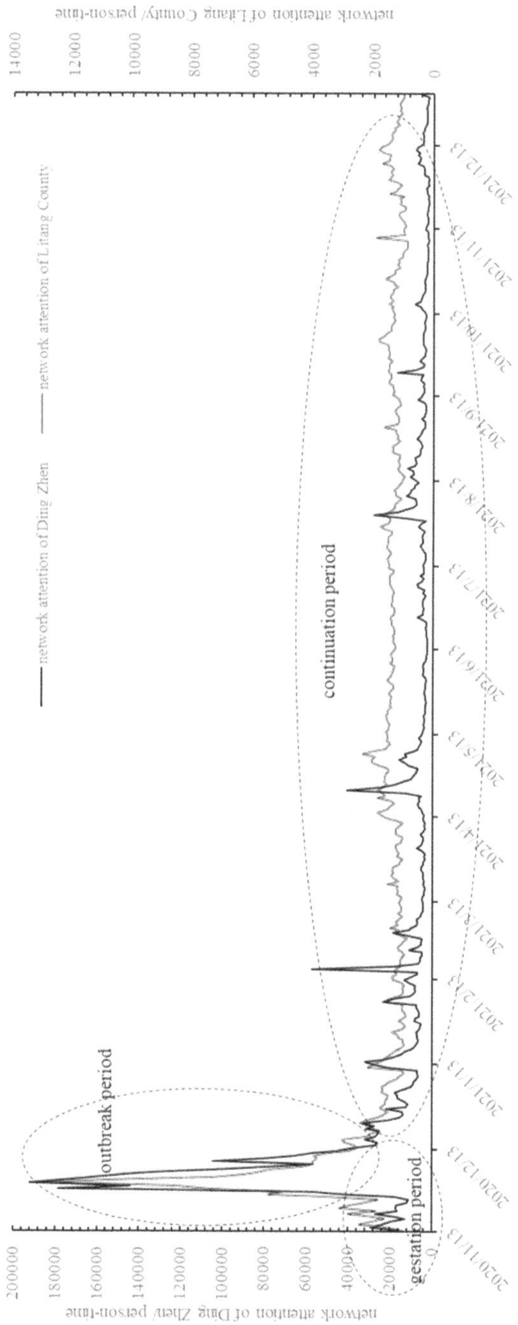

Fig. 1 Changes in Ding Zhen's and Litang County's national network attention in long term

short period of time, reaching a peak quickly, but at the same time, it also quickly returns to a relatively stable stage.

3.2 Analysis of Differences in Network Attention of Destination

3.2.1 Annual Difference

The total amount of network attention and year-on-year growth rate in Litang County from 2017 to 2019 are obtained through Baidu Index (Fig. 3). From Fig. 3, before 2020, the network attention of Litang County shows a slow decreasing trend, which drops by 3.85% in 2018 and 3.12% in 2019. From 2020, the network attention of Litang County shows an increasing trend, of which the growth rate exceeds 50% in 2020 and slows down in 2021, but still 20.76%. After 2020, most domestic tourist destinations suffered varying degrees of tourist loss due to the impact of the Covid-19 pandemic, but Litang County still achieved contrarian growth after 2020. Thus the endorsement has a significant impact on the network attention of Litang County, effectively enhancing the popularity of Litang County and promoting the tourism development here.

From the network attention of Litang County in each month from 2017 to 2021(Fig. 4), Litang County's network attention from 2017 to 2019 shows two peaks, and the second peak is lower than the first peak, so it presents an inverted U shape as a whole. Every year the first peak appears in July or August, the second peak appears in October, while the trough appears in January or February. In addition, there is a slow upward trend in January to July and September to October, and a downward trend in August to September and October to December. It can be seen that summer and autumn are peak tourist seasons in Litang County, while winter is the off-season tourism. The trend of network attention from January to October 2020 is basically the same as that of January to October 2017 to 2019, and due to the impact of the Covid-19 pandemic, the network attention of each month is lower than that of each month of 2017–2019. However, from November to December 2020, there is a surge in Litang County's network attention, mainly because Ding Zhen's popularity on the Internet leads to an increase in Litang County's network attention. Compared with the trend of network attention in 2020, network attention in 2021 is relatively stable, but it also fluctuates. After Ding Zhen's endorsement, the monthly network attention in Litang County in 2021 is much higher than that in 2017–2019, among which January 2021 is still in the stage of Ding Zhen's popularity, so the network attention is still at a relatively high level. After February 2021, Litang County's network attention appears three peaks, that is, May, July, and October, corresponding to the Labor Day Holiday, summer holiday, and National Day Holiday respectively.

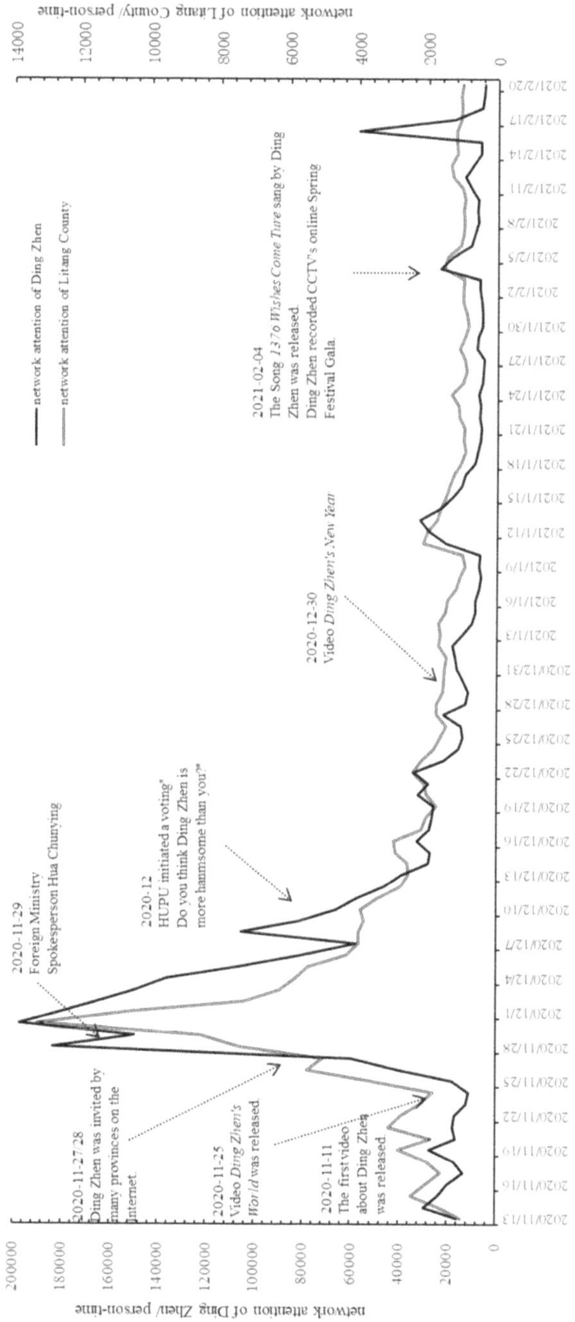

Fig. 2 Changes in Ding Zhen's and Litang County's national network attention in short term

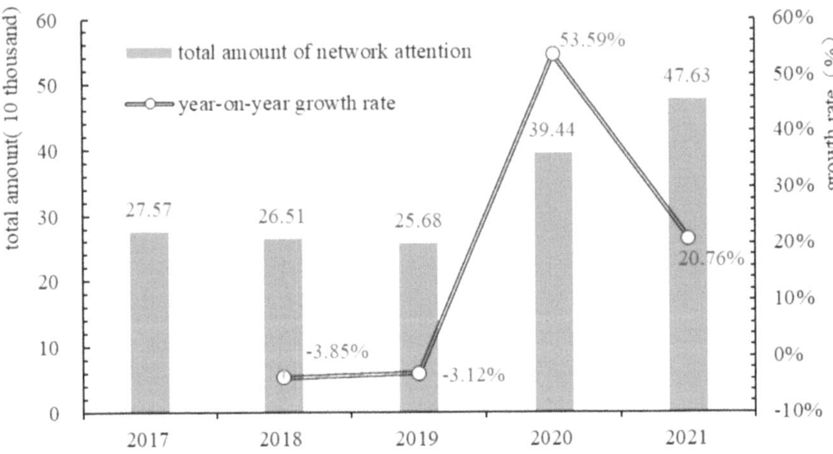

Fig. 3 Litang County's national network attention and growth rate (2017–2021)

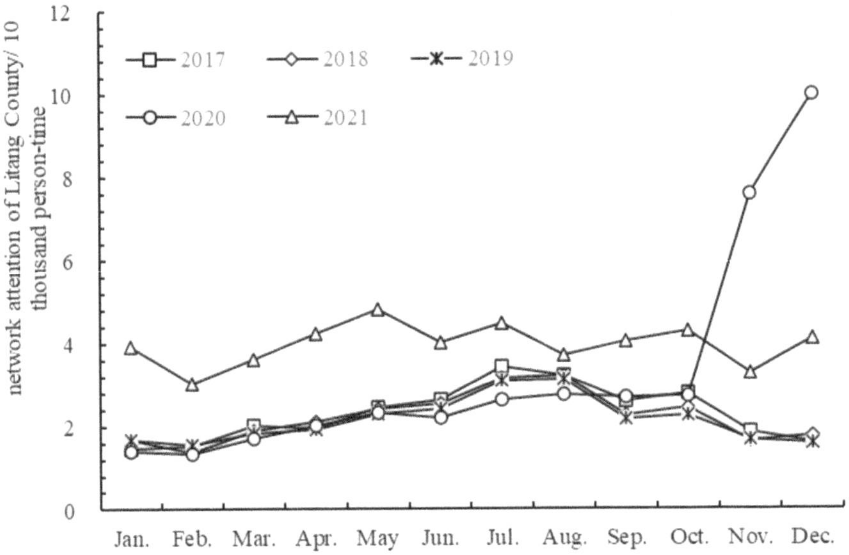

Fig. 4 Litang County's national network attention by month (2017–2021)

3.2.2 Seasonal Differences

The seasonal concentration index is used to measure the seasonal concentration degree of Litang County's network attention from 2017 to 2021, and the result of each year is 2.27, 2.10, 2.03, 6.49 and 1.01 (Table 1). As the result shows, the seasonal concentration index from 2017 to 2019 is all greater than 2, and shows a downward trend, indicating that the seasonal difference is large. In particular, the

Table 1 Monthly percentage and seasonal concentration index of Litang County's national network attention

Month	Year				
	2017	2018	2019	2020	2021
Jan	5.30	6.32	6.50	3.57	8.25
Feb	5.55	5.16	6.02	3.40	6.39
Mar	7.30	7.12	7.18	4.31	7.59
Apr	7.10	7.99	7.51	5.12	8.91
May	8.89	9.07	8.99	5.89	10.15
Jun	9.52	9.64	9.47	5.62	8.43
Jul	12.41	11.91	12.06	6.68	9.44
Aug	11.68	12.12	12.22	7.02	7.78
Sep	9.41	8.57	8.50	6.83	8.50
Oct	10.14	9.21	8.84	6.91	9.06
Nov	6.79	6.24	6.54	19.25	6.88
Dec	5.90	6.64	6.16	25.40	8.63
R	**2.27**	**2.10**	**2.03**	**6.49**	**1.01**

seasonal concentration index in 2020 has reached 6.49, mainly due to the popularity of Ding Zhen from November to December. However, compared with 2017 to 2020, the seasonal concentration index in 2021 is only 1.01, which reflects that seasonal difference in 2021 is quite small, indicating that to some extent, Ding Zhen's endorsement has narrowed the seasonal difference in Litang County's network attention.

3.2.3 Differences in Golden Week

The Labor Day Golden Week and the National Day Golden Week are two periods when people have high travel intentions, so there will also be high-intensity short-term tourism flows [28]. In order to explore the centralized distribution characteristics of Litang County's network attention during these two holidays, the study calculates the intra-week distribution skewness index during these two periods (Table 2). In addition, the Labor Day Holiday is usually 5 days, so the research time period is the Labor Day Holiday plus two days before and after the holiday [25]. Judging from the results, the intra-week distribution skewness index of Labor Day Golden Week fluctuates every year, indicating that the network attention in 2017, 2019, and 2021 concentrates in the later period, and in 2018 and 2020, the network attention concentrates in the early period. This might be because the climate and oxygen content of Litang County are still low in spring, and it is not suitable for many people to travel, so the skewness of network attention during the Labor Day Holiday is not stable. The intra-week distribution skewness index of the National Day Golden Week is less than 0, indicating that the network attention in Litang County is concentrated

Table 2 National network attention and intraweek distribution deviation index (T) of Litang County's during Labor Day and National Day

Date	Year				
	2017	2018	2019	2020	2021
Apr. 30	799	765	678	727	1799
May 1	696	817	730	794	2011
May 2	863	747	704	917	2011
May 3	762	664	739	844	2014
May 4	795	808	706	837	1928
May 5	986	680	846	775	2328
May 6	855	660	833	745	1970
T Value	**41.35**	**−35.95**	**46.72**	**−3.97**	**26.48**
Oct. 1	1172	1110	1061	1200	1514
Oct. 2	1467	1196	1126	1259	1792
Oct.3	1505	1042	1118	1445	1835
Oct. 4	1324	1113	963	1401	1655
Oct. 5	1088	879	901	1198	1549
Oct. 6	1002	985	769	1104	1403
Oct. 7	898	919	703	1007	1403
T Value	**−89.78**	**−55.95**	**−105.67**	**−46.16**	**−43.85**

in the early stage of the Golden Week, which is related to the fact that tourists would search for relevant information in advance to prepare for travel. In addition, intraweek distribution skewness index of the National Day Golden Week in 2020 and 2021 has increased compared with that in 2017–2019, indicating that tourists' network attention to Litang County gradually tends to even out from the initial concentrated distribution.

4 Spatial Distribution Characteristics of Network Attention

4.1 Spatial Distribution Consistency Analysis of Network Attention of Celebrity and Destination

In order to see the spatial distribution of network attention of Dingzhen and Litang county more intuitively, the study uses ArcGIS 10.6 to make distribution maps (Figs. 5 and 6) of network attention from November 13, 2020 to December 31, 2021. As the maps shows, Ding Zhen's network attention is generally higher in the east and lower in the west, among which the eastern coastal, southern coastal and Sichuan province has higher network attention. While the spatial distribution of Litang County's network attention is overall higher in the south and lower in the north, among which the eastern coastal, southern coastal and Sichuan province have higher network attention than other regions, and the northwestern and northeastern regions have relatively low network attention. Therefore, the distribution of network attention of Ding Zhen is consistent with that of Litang County. Using the natural breaks classification method, the network attention of 31 provinces could be divided into 5 levels from large to small respectively. It can be seen that Guangdong, Sichuan, Jiangsu, Zhejiang, and Beijing has a high level of network attention to Ding Zhen, and Sichuan is the province with the highest network attention, which might be because Litang County is in Sichuan Province and Guangdong, Jiangsu, Zhejiang and Beijing has relatively developed economy; Shanghai, Shandong, Henan, Hubei and other central provinces are at a medium level; Provinces in the northeastern and northwestern regions of China such as Heilongjiang, Inner Mongolia, and Xinjiang are at a relatively low level.

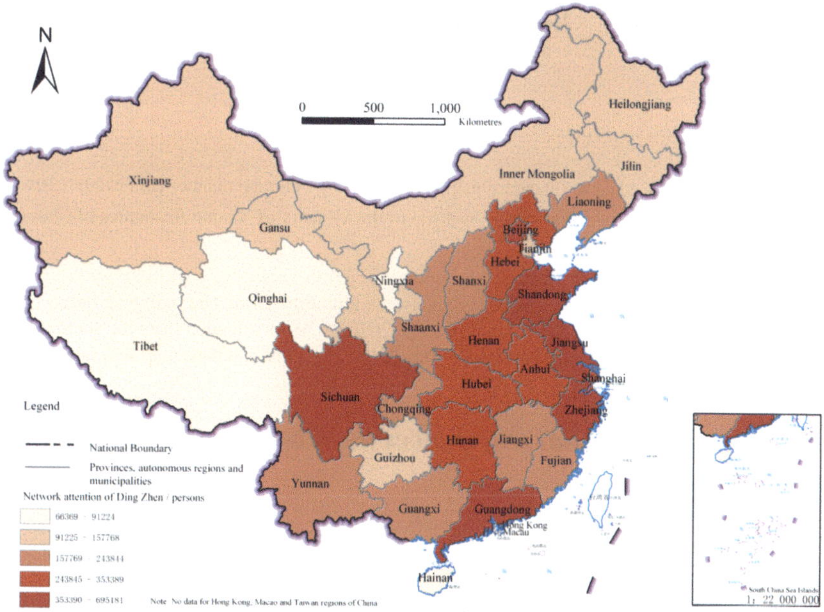

Note: The map was based on the standard map (1:11 000000) with map number GS (2020)4630 downloaded from the standard map service website of the Ministry of Natural Resources of China, and the base map was not modified.

Fig. 5 Distribution maps of Ding Zhen's network attention. *Note* The map was based on the standard map (1:11 000,000) with map number GS (2020)4630 downloaded from the standard map service website of the Ministry of Natural Resources of China, and the base map was not modified

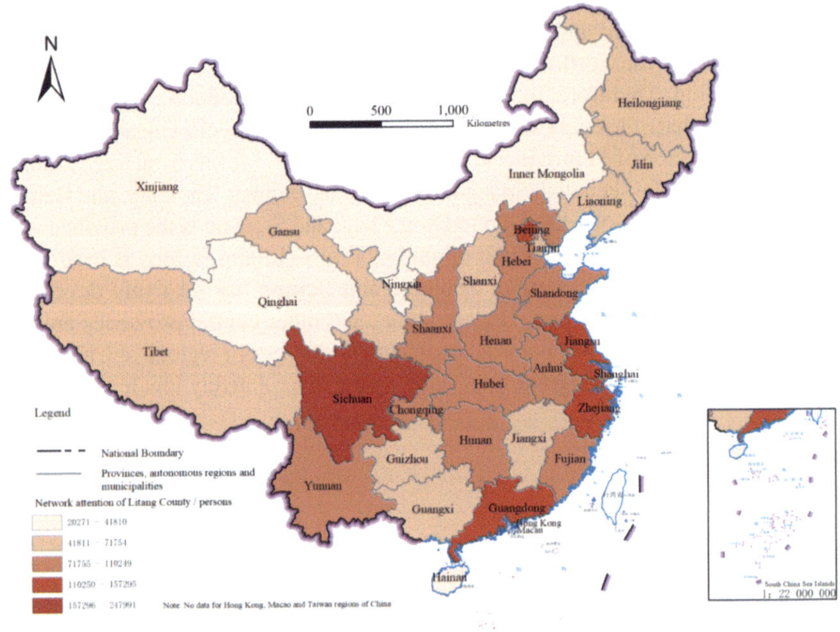

Note: The map was based on the standard map (1:11 000000) with map number GS (2020)4630 downloaded from the standard map service website of the Ministry of Natural Resources of China, and the base map was not modified.

Fig. 6 Distribution maps of Litang County's network attention. *Note* The map was based on the standard map (1:11 000,000) with map number GS (2020)4630 downloaded from the standard map service website of the Ministry of Natural Resources of China, and the base map was not modified

4.2 *Analysis of Spatial Differences in Network Attention of Destination*

4.2.1 *Inter–Provincial Difference*

Judging from the network attention garnered by Litang County and its growth rate across 31 provinces (Table 3), there are noticeable disparities in spatial distribution. Each province experienced a significant increase in network attention following the endorsement. Prior to and following Ding Zhen's endorsement, Sichuan, Guangdong, and Beijing consistently held the top three positions in terms of network attention towards Litang County. Chongqing, Zhejiang, Jiangsu, and Shanghai witnessed fluctuating levels of network attention but remained within the top ten rankings. Conversely, Inner Mongolia, Qinghai, Ningxia, Hainan, and Xinjiang consistently ranked near the bottom, occupying the fifth position from the bottom to the last.

Further analysis reveals that Sichuan, Chongqing, and Litang County are relatively proximate, while Hainan and Litang County are distant. Additionally, regions such as

Table 3 Network attention and growth rate of Litang County from all regions of the country (excluding Hong Kong, Macao and Taiwan) (2017–2021)

Region	2017	2018		2019		2020		2021	
	X_{2017}	X_{2018}	rate	X_{2019}	rate	X_{2020}	rate	X_{2021}	rate
Northeast China	**58,348**	**56,172**	**−3.73%**	**56,623**	**0.80%**	**93,266**	**64.71%**	**130,995**	**40.45%**
Heilongjiang	15,868	13,273	−16.35%	16,160	21.75%	33,389	106.62%	38,863	16.39%
Jilin	13,850	12,872	−7.06%	10,385	−19.32%	18,330	76.50%	36,917	101.40%
Liaoning	28,630	30,027	4.88%	30,078	0.17%	41,547	38.13%	55,215	32.90%
North China	**167,702**	**171,208**	**2.09%**	**165,932**	**−3.08%**	**227,486**	**37.10%**	**297,127**	**30.61%**
Beijing	73,450	72,248	−1.64%	68,081	−5.77%	92,129	35.32%	103,185	12.00%
Tianjing	21,149	24,991	18.17%	20,091	−19.61%	28,825	43.47%	46,458	61.17%
Shanxi	24,545	24,528	−0.07%	25,514	4.02%	34,680	35.93%	46,219	33.27%
Hebei	37,680	36,830	−2.26%	37,919	2.96%	52,881	39.46%	68,806	30.11%
Inner Mongoria	10,878	12,611	15.93%	14,327	13.61%	18,971	32.41%	32,459	71.10%
East China	**320,740**	**327,087**	**1.98%**	**303,620**	**−7.17%**	**459,017**	**51.18%**	**541,887**	**18.05%**
Shanghai	58,461	58,718	0.44%	54,260	−7.59%	76,866	41.66%	85,662	11.44%
Jiangsu	60,085	60,391	0.51%	57,744	−4.38%	82,089	42.16%	96,541	17.61%
Zhejiang	62,531	62,432	−0.16%	55,448	−11.19%	81,208	46.46%	100,273	23.48%
Anhui	30,350	35,273	16.22%	32,149	−8.86%	47,540	47.87%	62,635	31.75%
Fujian	37,377	37,588	0.56%	32,804	−12.73%	49,034	49.48%	63,590	29.69%
Jiangxi	26,053	23,932	−8.14%	23,693	−1.00%	42,706	80.25%	51,320	20.17%
Shandong	45,883	48,753	6.26%	47,522	−2.52%	79,574	67.45%	81,866	2.88%
South China	**114,521**	**117,825**	**2.89%**	**111,710**	**−5.19%**	**163,644**	**46.49%**	**204,468**	**24.95%**
Guangdong	78,422	81,989	4.55%	79,271	−3.32%	108,204	36.50%	121,720	12.49%
Guangxi	29,687	30,234	1.84%	27,048	−10.54%	40,511	49.77%	57,589	42.16%

(continued)

Table 3 (continued)

Region	2017	2018		2019		2020		2021	
	X_{2017}	X_{2018}	rate	X_{2019}	rate	X_{2020}	rate	X_{2021}	rate
Hainan	6412	5602	−12.63%	5391	−3.77%	14,929	176.92%	25,159	68.52%
Central China	**137,110**	**137,986**	**0.64%**	**128,739**	**−6.70%**	**178,884**	**38.95%**	**215,982**	**20.74%**
Henan	47,739	48,126	0.81%	45,079	−6.33%	63,148	40.08%	74,995	18.76%
Hubei	46,782	47,304	1.12%	44,366	−6.21%	61,978	39.70%	75,139	21.23%
Hunan	42,589	42,556	−0.08%	39,294	−7.67%	53,758	36.81%	65,848	22.49%
Region	2017	2018		2019		2020		2021	
	X_{2017}	X_{2018}	rate	X_{2019}	rate	X_{2020}	rate	X_{2021}	rate
Northwest China	**99,878**	**110,234**	**10.37%**	**104,012**	**−5.64%**	**137,210**	**31.92%**	**174,752**	**27.36%**
Shaanxi	45,954	48,007	4.47%	48,290	0.59%	56,697	17.41%	67,106	18.36%
Gansu	26,554	28,481	7.26%	25,730	−9.66%	31,685	23.14%	38,836	22.57%
Qinghai	15,187	17,551	15.57%	16,082	−8.37%	23,307	44.93%	25,737	10.43%
Ningxia	3836	6830	78.05%	4191	−38.64%	10,343	146.79%	14,005	35.41%
Xinjiang	8347	9365	12.20%	9719	3.78%	15,178	56.17%	29,068	91.51%
Southwest China	**343,503**	**343,596**	**0.03%**	**335,156**	**−2.46%**	**398,970**	**19.04%**	**437,036**	**9.54%**
Sichuan	165,831	162,359	−2.09%	160,404	−1.20%	182,948	14.05%	204,301	11.67%
Guizhou	32,164	33,091	2.88%	30,683	−7.28%	46,031	50.02%	47,434	3.05%
Yunnan	45,989	46,949	2.09%	46,534	−0.88%	54,388	16.88%	63,410	16.59%
Chongqing	64,340	62,208	−3.31%	59,593	−4.20%	71,511	20.00%	79,232	10.80%
Tibet	35,179	38,989	10.83%	37,942	−2.69%	44,092	16.21%	42,659	−3.25%

Note X_{2017}、X_{2018}、X_{2019}、X_{2020} and X_{2021} represent the network attention of each region in 2017, 2018, 2019, 2020 and 2021 respectively. The rate here is the year-on-year growth rate

Table 4 Gini coefficient of Litang County's network attention (2017–2021)

Year	Gini coefficient
2017	0.36
2018	0.36
2019	0.37
2020	0.32
2021	0.3

Guangdong, Beijing, Zhejiang, Jiangsu, and Shanghai exhibit relatively high levels of economic development. Hence, the spatial distribution appears to be influenced by factors such as spatial proximity, economic development, and population size. The growth rate of the network attention of Litang County in various regions was extremely low before Ding Zhen's endorsement, and even had a negative growth, but since 2020, the network attention of various regions to Litang County had achieved greater. In 2020, the highest growth rate reached 64.71% (Northeast region), and even the lowest was nearly 20% (Southwest region). Although the growth rate decreased in 2021, it still maintained a growth rate of more than 10% (except Southwest region). It can be seen that celebrity endorsements can awaken various regions' network attention to tourist destinations.

According to the formula, the Gini coefficient from 2017 to 2021 (Table 4) is in the range of 0.3–0.4. As Table 4 shows, Gini coefficient from 2017 to 2019 fluctuates between 0.36 and 0.38, and then decreases to 0.32 and 0.30 in 2020 and 2021 respectively, which decreases by 13.51% and 18.92% compared with 2019. It can be seen that from 2017 to 2021, the concentration degree of network attention to Litang County in all regions was at an above medium level, and the concentration degree decreased after Ding Zhen's endorsement. It indicates that the network attention in Litang County is relatively concentrated in spatial distribution, but the concentration degree is not high, and Ding Zhen's endorsement has reduced the spatial distribution of agglomeration.

4.2.2 Regional Difference

From Table 5, from 2017 to 2019, the annual coefficient of variation in the seven regions shows little difference and small fluctuation. Among the seven regions, the coefficient of variation in South China, Northwest China, Southwest China and North China are large, all of which are more than 50%, indicating that there is a big difference in network attention within the region. Especially, the coefficient of variation in South China in 2018 and 2019 reaches 0.81 and 0.83, and that is because Guangdong has a very high network attention in South China (ranking second) while Hainan has a very low network attention (ranking first or second from the bottom). However, the coefficient of variation in Central China, East China and North China is small, indicating that there is little difference in network attention within the region. Especially, the coefficient of variation in Central China is only 0.05, 0.05 and 0.06, which

Table 5 Coefficient of variation of Litang County's network attention (2017–2021)

	2017	2018	2019	2020	2021
Northeast China	0.34	0.43	0.44	0.31	0.19
North China	0.65	0.6	0.58	0.57	0.42
East China	0.3	0.29	0.29	0.26	0.22
South China	0.79	0.81	0.83	0.72	0.59
Central China	0.05	0.05	0.06	0.07	0.06
Northwest China	0.75	0.68	0.75	0.6	0.51
Southwest China	0.73	0.7	0.71	0.66	0.68
Total	**0.13**	**0.13**	**0.13**	**0.11**	**0.1**

is related to the similar network attention among provinces in Central China. In 2020 and 2021, by contrast, the variation coefficient of large area all decreased, but the coefficient of variation in South China, Northwest China, Southwest China, North China was still larger than that in Central China, East China and North China. It indicates that the endorsement of Ding Zhen promoted the network attention of each region and narrowed the difference of network attention among different provinces in different regions. The coefficient of variation from 2017 to 2019 was 0.13, which is small, indicating that there is a certain difference in network attention between provinces, but the difference is small. Moreover, the coefficient of variation in 2020 and 2021 decreases by 15.09% and 23.08% compared with 2019, which shows that Ding Zhen's endorsement has reduced the difference of network attention between provinces.

5 Conclusions and Suggestions

5.1 Conclusions

Using Dingzhen's endorsement of Litang County as a case study, this paper examines the network attention towards Litang County from 2017 to 2021 and Dingzhen's own network attention from 2020 to 2021. Through the application of quantitative indicators, the paper analyzes the distribution characteristics of network attention towards tourist destinations before and after celebrity endorsement, considering both temporal and spatial perspectives. The primary conclusions are outlined below:

There are both temporal synchronization and spatial difference consistency in network attention of celebrity and network attention of tourist destination. In time dimension, the changes of network attention of celebrities and tourist destinations were synchronized. With the support of celebrity effect, the changes of both shows a unimodal trend, which could be divided into gestation period, outbreak period

and continuation period. In space dimension, the spatial distribution of network attention of celebrities and tourist destinations is consistent, showing a pattern of high in the south and low in the north as well as high in the east and low in the west. Apart from the eastern and southern coastal areas, provinces where tourist destination are located also had high network attention.
In terms of temporal evolution, after celebrity endorsement, the annual network attention of tourist destination increases rapidly, and the growth rate turns from negative to positive. The seasonal difference is reduced, and June to October is the peak season for tourism, and November to February the next year is the off-season for tourism. Distribution skewness of network attention on Labor Day Golden Week is unstable, and that on National Day Golden Week Changed from concentrating in the previous week to distributing evenly.
In terms of spatial evolution, after celebrity endorsement, the network attention of each region increases significantly, and the growth rate changes from negative to positive. The degree of agglomeration is above medium, and the degree of agglomeration decreases after the celebrity endorsement. The distribution differences in South China, Southwest and Northwest regions are large, and the differences between regions and provinces are reduced after celebrity endorsement.

5.2 *Suggestions*

Celebrity endorsements can undoubtedly boost the visibility of tourist destinations and attract tourists' attention, yet increased popularity doesn't necessarily equate to enhanced reputation and loyalty. Building upon this premise, the following suggestions are proposed:

Firstly, if a government of destination adopts a policy of celebrity endorsement, then the first step is to seize the initiative of celebrity endorsement. Litang County's swift response in promptly signing a contract with Ding Zhen following the initial video's viral circulation ensured that Ding Zhen's image remained untarnished by excessive commercialization, while also retaining control over the endorsement process. Moreover, by establishing social media accounts under the moniker "Litang Dingzhen" and consistently updating content across platforms like Tiktok, Bilibili, and Weibo, both Ding Zhen and Litang County garnered increased attention from a wider audience.

Secondly, the governments should capitalize on opportunities for tourism development and enhance supporting services. While celebrity endorsements may initially attract potential tourists, the lasting satisfaction, reputation, and loyalty of visitors hinge upon their on-site travel experiences and the quality of local services. Therefore, there's a pressing need to elevate operational standards, bolster supporting infrastructure, and enrich service offerings. Investments in upgraded tourist service centers, enhanced rest areas, and diverse festival activities can contribute to an exceptional tourism experience, fostering heightened satisfaction, reputation, and loyalty among tourists.

Thirdly, government should prepare to mitigate the potential negative impacts of celebrity endorsements. While celebrity endorsements can undoubtedly drive tourist traffic to a destination, they also carry the risk of alienating tourists if the endorsed celebrity's actions or statements are deemed inappropriate. Thus, destinations must exercise caution in selecting celebrities for endorsement and be prepared to implement strategic public relations management measures to navigate any potential controversies or fallout.

5.3 Future Research

The study contributes to the existing research on celebrity endorsement of destinations; however, it is not without its limitations. Firstly, the data range selected for this research spans from January 1, 2017, to December 31, 2021. While this timeframe encompasses five years of network attention data, it fails to capture the longer-term impact of Dingzhen's online popularity, which began in November 2020, on the tourism development of Litang County. Secondly, due to constraints in data collection, the underlying factors driving spatial distribution differences in network attention were not explored. Thirdly, the use of network attention data from the Baidu search engine may introduce bias, as some individuals may not utilize this specific search engine. Consequently, further investigation is warranted to confirm the actual customer influx into Litang County following Ding Zhen's endorsement.

Several avenues for future research emerge. Firstly, examining the enduring effects of celebrity endorsements over an extended period can provide deeper insights. Secondly, delving into the influencing factors shaping the spatial and temporal distribution of network attention warrants further exploration. Lastly, assessing the efficacy and unique characteristics of celebrity endorsements across different types of celebrities endorsing destinations holds promise for future investigation.

Acknowledgement The research was supported by the National Natural Science Foundation of China (Grant No. 42271238), the National Natural Science Foundation of Guangdong Province of China (Grant No.2021A1515011627), and Guangzhou Humanities and Social Sciences Key Research Base (ZKXM202135).

References

1. Liu, Z., Yang, Y., Sui, X.: Internet development, market vitality and tourism economic growth: Based on the perspective of spatial spillover. Tourism Science 1–25 (2022)
2. Wu, Z., Peng, A., Zhang, M.: Influence of "celebrity effect" on international inbound tourism: Evidence from Guilin. Res. Develop. Market **38**(04), 499–505 (2022)
3. Mccracken, G.: Who is the celebrity endorser? Cultural foundations of the endorsement process. J. Consum. Res. **16**(3), 310–321 (1989)

4. Knoll, J., Matthes, J.: The effectiveness of celebrity endorsements: a meta-analysis. J. Acad. Mark. Sci. **45**(1), 55–75 (2017)
5. Hedhli, K.E., Zourrig, H., Becheur, I.: Celebrity endorsements: investigating the interactive effects of internalization, identification and product type on consumers' attitudes and intentions. J. Retail. Consum. Serv. **58**, 102260 (2021)
6. Halder, D., Pradhan, D., Chaudhuri, H.R.: Forty-five years of celebrity credibility and endorsement literature: review and learnings. J. Bus. Res. **125**, 397–415 (2021)
7. Park, H.J., Lin, L.M.: The effects of match-ups on the consumer attitudes toward internet celebrities and their live streaming contents in the context of product endorsement. J. Retail. Consum. Serv. **52**, 101934 (2020)
8. Albert, N., Ambroise, L., Valette-Florence, P.: Consumer, brand, celebrity: which congruency produces effective celebrity endorsements? J. Bus. Res. **81**, 96–106 (2017)
9. Ohanian, R.: Construction and validation of a scale to measure celebrity endorsers' perceived expertise, trustworthiness, and attractiveness. J. Advert. **19**(3), 39–52 (1990)
10. Zhang, H., Xu, H., Gursoy, D.: The effect of celebrity endorsement on destination brand love: a comparison of previous visitors and potential tourists. J. Destin. Mark. Manag. **17**, 100454 (2020)
11. Zhou, X.: The Influencing factors and the theoretical discussion of the communication effect of celebrity advertising. Adv. Psychol. Sci. **17**(04), 811–820 (2009)
12. Zhang, H., Xu, H., Li, Y.: Celebrity endorsement's influence on destination brand love: combining the perspectives of source credibility and the match-up hypothesis. Tourism Tribune **36**(09), 60–74 (2021)
13. Mccormick, K.: Celebrity endorsements: influence of a product-endorser match on millennials attitudes and purchase intentions. J. Retail. Consum. Serv. **32**, 39–45 (2016)
14. Yu, S., Hu, Y.: When luxury brands meet China: the effect of localized celebrity endorsements in social media marketing. J. Retail. Consum. Serv. **54**, 102010 (2020)
15. Shen, X., Li, T., Qu, Y.: The Influence of celebrity endorsement on tourism destination brand equity: a perspective of celebrity credibility. Econom. Manage. **38**(04), 138–148 (2016)
16. Liu, L.: Study on the influential mechanism of celebrity endorsing tourist destination: an application of identification theory. Res. Econom. Manage. **10**, 16–20 (2016)
17. Liu, H., Yan, Q., Yang, Y.: Celebrity endorsement on impact of tourist destination brand value analysis: take a tourist attraction in an ancient town in East China as an example. Inquiry Econom. Issues **07**, 87–92 (2013)
18. Roy, S., Dryl, W., de Araujo Gil, L.: Celebrity endorsements in destination marketing: a three country investigation. Tour. Manage. **83**, 104213 (2021)
19. Zhu, F., Xiang, Y., Ding, X.: The impact of Internet on the tourist destination evolution: a case study of Wuyuan. Hum. Geogr. **28**(05), 147–152 (2013)
20. Ma, L., Ma, M.: Influence of "Tianjiaxia" events to Qingdao tourism network attention. Res. Develop. Market **34**(01), 83–87 (2018)
21. Lin, Z., Ma, Y., Liu, X., et al.: Spatial and temporal features of network attention of science areas. Res. Sci. **34**(12), 2427–2433 (2012)
22. Ma, L., Sun, G., Huang, Y., et al.: A correlative analysis on the relationship between domestic tourists and network attention. Econ. Geogr. **31**(04), 680–685 (2011)
23. Long, M., Sun, G., Ma, L., et al.: An analysis on the variation between the degree of consumer attention of travel network and tourist flow in regional tourism: a case of Sichuan Province. Areal Res. Develop. **30**(03), 93–97 (2011)
24. Ma, L., Guo, L., Wu, Zi.: The influence of *Daddy, Where're We Going?* to its locations' tourism: based on the analysis of tourists' network attentions. Tourism Sci. 30(03), 57–68 (2016)
25. Zhu, D., Li, L., Li, X., et al.: Spatial-temporal differences and influencing factors of network attention of popular scenic spots in Xinjiang. J. Northwest Normal Univ. (Nat. Sci.) **57**(06), 110–117 (2021)
26. Long, M., Ma, L.: Spatial distribution differences of star hotels in China and their influencing factors. Econom. Forum **08**, 87–92 (2013)

27. Hao, L., Yang, F., Zhang, Q.: Rediscussion on statistical measure on Gini coefficient. Statist. Decision **37**(07), 27–32 (2021)
28. Wang, Q., Cao, W., Zhang, L.: Study on spatial-temporal distribution of red tourism network in Anhui Province. Res. Develop. Market 1–11 (2022)
29. Mao, T., Jiang, G., Li, Y., et al.: Emotion communication characteristics of network hot events in the new media age. Inform. Sci. **37**(04), 29–35 (2019)

Measuring the Tourists' Connection to Nature in Chinese National Wetland Parks

Zhihong Liu and Jin Yin

Abstract The purpose of this study is to develop a scale for measuring the tourists' connection to nature in the context of Chinese national wetland park. Following a rigorous process of scale development, a three-dimension (*natural identity, emotional affinity, and natural dependence*) model of connection to nature is established. The study develops 12 measurement items to construct a survey instrument of tourists' connection to nature. To test the predictive validity of the scale, pro-environmental behavior is measured as an outcome variable of tourists' connection to nature. The study tries to provide a framework for future empirical investigations in nature-based tourism settings, and offer a useful tool for such purpose.

Keywords Tourists' connection to nature · Pro-environmental behavior · Xixi National Wetland Park · Scale development

1 Introduction

The unprecedented COVID-19 pandemic swept the globe from 2019 to 2023, affecting almost all aspects of social life. The pandemic has caused a huge public mental health crisis [1]. Fear of illness, social isolation, and reduced income all contributed to the high incidence of psychological problems [2]. High prevalence of acute stress, anxiety and depression symptoms were observed during the epidemic [3], which could be a risk factor for post-traumatic stress disorder and other mental disorders after the outbreak. The natural environment is less threatening and makes people feel safe. Spending time in nature and appreciating nature yields many benefits for people such as recovering from stress and fatigue, promoting mental and physical health [4, 5]. However, social media and digital technology impedes people's

Z. Liu (✉)
School of Culture Tourism, Guangdong University of Finance & Economics, Guangzhou, China
e-mail: liuzhihong1130@gdufe.edu.cn

J. Yin
Department of Tourism Management, School of Humanities, Southeast University, Nanjing, China
e-mail: allen-yin@seu.edu.cn

Y. Luo et al. (eds.), *Tourist Experience and Perception in the Pandemic and Post-Pandemic Era*, https://doi.org/10.1007/978-981-96-1820-0_5

accessing of nature, humans tend to experience nature artificially rather than directly, thus creating a distance from nature [6]. Distance from nature may lead to a decrease in human well-being [7]. In environmental psychology, a great number of studies have shifted from local and specific environmental problems, such as energy use, littering and material recycling to broader conceptualizations of human–nature relationship [8]. Green operations which connect people with nature are increasingly being applied to improve individual's mental health [9]. Indeed, the human–nature relationship is reciprocal, human behavior affects the well-being of nature, and changes in the natural environment in turn affect the well-being of humans. The benefits of connection to nature are strongly related to the awareness of environmental issues, green behaviors, and human well-being [10].

Environmental problems such as global climate change and resource depletion have been increasingly drawing the attention of ecologists and social psychologists. Despite recognizing that human behaviors are the cause of many environmental problems, individuals usually fail to choose behaviors that will minimize the negative impacts on the environment [11]. There is a gap between individuals' attitudes about environmental issues and their behaviors. It is necessary to find ways to address this gap and transform environment concerns into pro-environmental behaviors [12]. Researches have pointed out that a critical aspect of mitigating the environmental crisis is reconnecting humans to nature and establishing harmonious relationship between human and nature [13].

Relationship between human and nature is a long-standing theme in ecology and eco-psychology [14]. The biophilia hypothesis proposes that humans are physiologically depending on nature and affiliated with nature [15]. Nature is a community to which humans belong, human and nature are interdependent rather than separate. Leopold [14] has pointed out that we abuse land because we regard it as a commodity belonging to us, when we see land as a community to which we belong, we may begin to use it with love and respect. When the sense of self is expanded to include the natural community, environmental destruction behavior will be considered as self-destruction [16]. Phenomenal descriptions and empirical studies both indicate that human's connection to nature is an important factor for fostering pro-environmental behavior. Confronted with severe environmental issues, what modern people need to do is to rebuild the connection between human and nature. Over the last two decades, there has been an increasing number of empirical studies on human–nature connection [5, 8, 17]. A number of concepts related to human's connection to nature has been put forward, such as emotional affinity toward nature [18], environmental identity [4], connectedness to nature [8], and love and care for nature [19]. Measurement scales of these concepts have been developed in different settings, including office buildings, controlled laboratory environment, and classrooms. Despite their apparent similarity, these concepts and measurement scales are often studied in isolation [20]. Thus, it is essential to conduct an overview and comparison towards the various related concepts in one study.

Connection to nature refers to the extent to which one person connects to nature and belongs to a part of the natural community [8, 21]. Although connection to nature has emerged as a widely used term in western environmental psychology, a clear conceptualization and measurement of the construct in the context of non-western culture is still in absence. The differences on ideology and cultural values across different culture backgrounds lead to the divergence of people's connection to nature. Social psychological studies have indicated that self-construal varies across cultures [22]. Based on the study of Singelis [23], "independent self-construal" in western cultures and "interdependent self-construal" in Chinese culture make the different way that individuals conceptualize their relationships with the natural world. Moreover, tourism development in China draws much from traditional Chinese culture, such as Confucianism, Taoism and the principle of harmony [24]. Harmony is the highest principle embedded in Chinese philosophical thinking, and a fundamental ideology for Chinese people to see the relationship between human and the natural world [25]. Chinese way of thinking includes human and nature in an integrated and united relationship, while western paradigm for tourism in natural sites follows a sustainable biocentric model, in which humans are not included in nature but separated from it [26, 27]. Therefore, Chinese understanding on the relationship between human and nature is different in many ways from western perspective. The above concepts and measurement scales related to connection to nature are mainly developed in western context and based on the ways of how westerners interact with nature, tourists' connection to nature in non-western culture like China has rarely been empirically examined. The present study thus aims to provide the needed empirical clarification of tourists' connection to nature and its measurement under the context of Chinese national wetland park.

2 Theoretical Background

2.1 Conceptualization and Measure of Connection to Nature

The term connection to nature originally refers to "ecological self" in eco-psychology [28], which means expanding human's identity to include the reciprocity between human and nature. In the last two decades, environmental psychologists have contributed to conceptualization and measurement of connection to nature. Previous studies have proposed various related concepts to express the feelings of connection with the surrounding natural world. Kals et al. [18] first introduce the concept of emotional affinity toward nature, which including various inclinations toward nature such as the love of nature. They argue that emotional affinity toward nature emphasizing the role of emotions in connection to nature, which is different from the cognitive beliefs about natural world. A unidimensional 16-item scale is developed and emotional affinity toward nature is empirically proved to predict the nature-protective behavior. Mayer and Frantz [8] propose the concept of connectedness to nature and develop a 14-item scale. They claim that connectedness to nature measures human's emotional feeling in community with nature and provides prediction of subjective well-being and ecological behavior. However, the nature of the concept is questioned by Perrin and Benassi [29], who suggest that connectedness to nature actually measures cognitive beliefs instead of emotional connections. Love and care for nature is another related concept concerned with affective aspect, which refers to people's clear recognition of nature value and personal sense of responsibility to protect nature. Perkins [19] develops a 15-item scale to measure love and care for nature and indicates that love and care for nature is strongly related to connectedness to nature.

From a cognitive perspective, Schultz [30] introduces the concept of inclusion of nature in the self, which is defined as the extent to which people include nature within their cognitive representation of self. A single-item graphical measure is developed to measure inclusion of nature in the self, including a choice of seven pairs of circles (one labeles nature and the other labeles self), ranging from entirely separate to nearly entirely overlapping. Later, Schultz et al. [21] argue the connection between human and nature is implicit and propose the concept of implicit connections with nature to assess the extent to which a person believes that he or she is part of nature. A modified implicit association test is employed by Schultz et al. The measure does not rely on respondent's self-report, however, it does not always correlate with explicit self-reporting measure of attitude.

Later researches conceptualize connection to nature from a multidimensional perspective. Clayton [4] proposes the concept of environmental identity, which assessing the extent to which the natural environment plays an important part in people's self-concept. Theoretically, environmental identity includes five dimensions: salience of interacting with nature, self-identification, ideology, positive emotions and autobiographical component of interacting with nature. Clayton develops a 24-item scale to measure environmental identity, but only finds a single factor in empirical study. Another multidimensional concept is nature relatedness (NR), which assessing individual's understanding of and appreciation for the connection with all other living things. According to Nisbet et al. [31], nature relatedness captures one's cognitive, affective and physical relationship with the natural world. Their empirical results indicated that both one-dimension structure and three-dimension (NR-Self, NR-Perspective, NR-Experience) structure were feasible. As a few items loaded on multiple dimensions, and the factors are strongly correlated, the one-dimension structure, 21-item scale seems to be more promising [31].

Interdependence theory points out that greater dependence on a relationship partner would result in greater commitment. With reference to the theory, Davis et al. [32] introduce an 11-item scale to measure commitment to the natural environment, which refers to the psychological attachment and long-term orientation toward nature. Connectivity with nature is another related concept put forward by Dutcher et al. [33]. It describes people's subjective perception of sameness between oneself, others, and nature. Dutcher et al. use a 4-item scale and one graphical measure of inclusion of nature in the self to measure connectivity with nature, however, the reliability of the scale is relatively low.

Although the above concepts place varying emphasis on different aspects of connection to nature (cognitive beliefs, affective affiliation, or relationship commitment), some common attributes could be drawn from these studies. A few studies summarize the potential overlaps between the different concepts and measurements. First, all the concepts elaborate the relationship between human and nature. Second, similar measurement items are used to measure different concepts. For instance, "a sense of oneness with nature" and "a connection with nature is important to me" are used in scales of different concepts. The measures of connectivity with nature and inclusion of nature in the self both employ the same graphical item. Third, empirical results indicate the convergence of different concepts. For example, Mayer and Frantz [8] find that connectedness to nature is strongly correlated with inclusion of nature in the self; Perkins [19] indicates that love and care for nature is strongly correlated to connectedness to nature and inclusion of nature in the self; Davis et al. [34] report that commitment to the natural environment is highly correlated with inclusion of nature in the self, connectedness to nature, and environmental identity; Tam [20] demonstrates that the measures of seven concepts related to connection to nature are strongly inter-correlated and converge to one single factor. Therefore, to facilitate a better understanding of connection to nature, integrating and clarifying the various concepts and measures in the existing studies is needed.

Encounters with nature can transform a person through the sense of connection, however, connection to nature may vary depending on the context in which it is tested [21]. The measurement of connection to nature in previous studies are mainly taken in classrooms, controlled laboratory environment, and offices, tourism setting is seldom examined. More importantly, culture has fundamental influences on an individual's self-construal, the way people conceptualize their relationship with nature varies in different cultures [22]. In western cultures, people hold the independent self-construal, and have a narrow view of self which emphasizes on the uniqueness and separateness of individuals. While in non-western cultures like China, people hold the interdependent self-construal, and self is defined more broadly and interconnected with other people and the social context [23]. As indicated by Schultz et al. [21], connection to nature is an extension of an interdependent self, which depends not only on the relationship with other people, but also the relationship with the natural environment around the person. Most of the applied measurements related to connection to nature are originated in the western context, which may not adapt to non-western culture and context. Chinese culture emphasizes unity and immanence, which leads to a harmony between man and nature [35]. The doctrine of "the Unity of Heaven and Humanity" is one of the most fundamental philosophical thoughts in the traditional Chinese way of thinking [36]. It is thus distinct from western perspective that separates human and nature, and considers nature as free from human interventions and artificiality [37]. The above differences will be incarnated in the human–nature relationships among Chinese tourists. In the present study, tourists' connection to nature refers to the extent to which a tourist connects to nature and belongs to a part of the natural community. The growing interest of tourists' connection to nature in China requires the development of validated measurements to examine tourists' relationship with nature.

2.2 *Connection to Nature in Tourism Study*

Regarding connection to nature and environmental behavior, a number of studies have demonstrated significant associations between these two constructs across different contexts including classrooms, natural settings, and farming regions [20, 33, 38]. For instance, Stewart and Craig [39] suggest that compared with frequent experience of a constructed environment, frequent experience of natural environment is more positively related to pro-environmental behavior. Davis et al. [32] report that higher levels of commitment to the environment predict higher levels of pro-environmental behavior. It has been argued that if people see themselves as a part of nature, they would not harm nature [8]. Love and care for nature is also found to be significantly linked to pro-environmental behavior [19]. Feeling a sense of strong connection to nature is an important antecedent of fostering the donating behavior of tourists [38]. However, the above relationships are not consistently reported through different studies. In the study of Schultz et al. [21], no significant association is found between implicit connections with nature and self-reported environmental behavior. Therefore, the literature has indicated connection to nature can motivate individuals to engage in pro-environmental behaviors, further research is also needed to test the relationship between individual's connection to nature and pro-environmental behavior in specific settings.

3 Methods and Results

3.1 *Initial Item Pool Generation*

The mixed method approach employing both qualitative and quantitative methods is adopted in the current study. A review of the literature and a few Chinese ancient classic books (e.g. The Book of Changes, Tao Te Ching, and Zhuangzi's view of nature) identifies the conceptual definitions of tourists' connection to nature, and further leads to the generation of an initial 43 items. In-depth interview is then employed to explore new aspects of tourists' connection to nature. Interviews proceeds until the information is deemed saturated. The researchers interviewed 25 tourists using open-ended questions. For example, recall your most recent experience in wetland parks, how do you evaluate the connection between you and nature? In reviewing the answers, we find that items generated from the interview are largely consistent with those in the literature review. Three new items "I feel a sense of awe for nature", "Human beings are part of nature", and "Being in nature, I can relax myself" are added. In addition, two rounds of focus group discussions with 12 participants were further conducted as a triangulation effort to assess the content validity of the items. Different wordings that share the same meanings are merged together. For example, "I am a part of nature, not separate from it" and "I think of the natural world as a community to which I belong". By combining the items generated from the literature review, in-depth interview and focus group, an initial set of 21 items to measure tourists' connection to nature is generated (Table 1).

3.2 Data Collection and Item Purification

A pilot study was conducted among tourists who visiting Xixi National Wetland Park. Participation in this survey is voluntary. After explaining the purpose of the survey, respondents are asked to fill in the self-administered questionnaire. A 5-point Likert scale ranging from *strongly disagree* (1) to *strongly agree* (5) is used to rate each item of tourists' connection to nature. A total of 150 questionnaires were distributed and 115 were returned, with an overall response rate of 76.7%. Internal consistency is measured using Cronbach's alpha. The reliability of the 21 items is 0.89, indicating that the 21-item scale was highly reliable. Following Kim et al. [40], four items that are poorly correlated with the total score ($r < 0.40$) are removed. This process results in 17 items being retained.

To refine the scale, a large-scale field survey was conducted in Xixi National Wetland Park. The questionnaire includes three main parts. The first part involves 17 items regarding tourists' connection to nature, the second part is designed to measure personal norms and pro-environmental behavior. Eight items on pro-environmental behavior are derived from Cheng et al. [41]. A 5-point Likert-type scale (from 1 = "strongly disagree" to 5 = "strongly agree") is employed for all items in the first and second parts. The third part includes tourists' demographic variables such as gender, age, occupation, monthly income and education level.

A convenience sampling approach is adopted in this study. Data were collected at four locations within the wetland park, namely the Hezhu Street, Cruise Terminal, Jiangcun Fair, and Hong-clan Ancestral Hall. These four sites are often crowed with tourists and thus convenient for data collection. Data were collected from 900 respondents during both weekends and weekdays, which included a range of visitors in the sample. An average of 180 copies were collected from each site, representing an overall response rate of 80%. Fifty-four questionnaires were removed because of missing data. This resulted in 666 questionnaires for the analysis. The female participants outnumber their male counterparts (53.5% vs 46.5%). The majority of respondents are in the age group of 26–35 (39.3), followed by the age group of 18–25 (37.8%). In terms of occupation, the largest group of respondents (30.3%) is enterprise staff, followed by the enrolled students (26.1%) and government and institution employees (24.6%). Regarding the education background, 52.9% of the respondents has completed an undergraduate degree, 20.0% are junior college graduates, and 12.5% are at postgraduate level.

The whole sample is randomly split into two subsamples using SPSS random case selection: one as calibration sample ($n = 331$) and the other as validation sample ($n = 335$). Exploratory factor analysis (EFA) is first adopted to identify underlying dimensions of tourists' connection to nature with the calibration sample. Principal component analysis with VARIMAX rotation is conducted with the 17 items. The

Table 1 Measurement items of tourists' connection to nature

Items	Sources
1. Human and nature are interdependent	Davis et al. [32]; In-depth interview and focus group
2. I am a part of nature, not separate from it	Clayton [4]; Nisbet et al. [31]; In-depth interview and focus group
3. I feel a close connection with the environment	Kals et al. [18]; In-depth interview and focus group
4. I feel a sense of oneness with nature when I spend time in the wetland park	Mayer & Frantz [8]; Perkins [19]; In-depth interview and focus group
5. Man and nature should coexist harmoniously	In-depth interview and focus group
6. Humans have the right to use natural resources any way we want (reversed)	Nisbet et al. [31]; In-depth interview and focus group
7. Animals and plants have the same rights with humans	Nisbet et al. [31]; In-depth interview
8. Humans have a lot in common with other species	Clayton [4]
9. When I spend time in nature, I can relax myself	In-depth interview and focus group
10. When I spend time in nature, I feel happy and satisfied	Kals et al. [18]; In-depth interview and focus group
11. When I spend time in nature, a feeling like security arises	Kals et al. [18]; In-depth interview and focus group
12. When I spend time in nature, I have a pleasant feeling of intimacy	Kals et al. [18]; In-depth interview and focus group
13. When I spend time in nature, I feel a sense of awe for nature	In-depth interview and focus group
14. Being a part of nature is an important part of who I am	Clayton [4]; In-depth interview and focus group
15. Connection to the environment in wetland is important to me	Davis et al. [32]; In-depth interview and focus group
16. I would feel that an important part of my life was missing if I could not enjoy nature from time to time	Clayton [4]; In-depth interview and focus group
17. I can derive spiritual sustenance from nature	Clayton [4]; In-depth interview and focus group
18. My personal welfare is independent of the welfare of the natural world	Mayer & Frantz [8]; In-depth interview and focus group
19. Behaving responsibly toward the Earth is part of my moral code	Clayton [4]
20. I need to be in the natural setting as much as possible	Perkins [19]; In-depth interview and focus group
21. I spend a lot of time in natural settings	Clayton [4]; In-depth interview and focus group

Kaiser–Meyer–Olkin (KMO) statistic (0.915) and Bartlett's test of sphericity (χ^2 = 1650.114; *df* = 105, p = 0.000) confirm sampling adequacy to run factor analysis. Five measurement items are removed due to low factor loading (<0.50) and low communalities (<0.40). The remaining 12 items are subject to another round of EFA analysis (KMO = 0.899, Bartlett's test of sphericity = 1367.034; *df* = 66; p = 0.000). Three factors are generated from the 12 items on the basis of the following criteria [42]: (1) eigenvalues greater than one; (2) screen plot of eigenvalues; (3) factor loadings greater than 0.50; (4) meaningfulness of the factors. The three factors account for 60.418% of the total variance of the items. The overall Cronbach's alpha of tourists' connection to nature is 0.866. The Cronbach's alphas for the three factors are 0.681, 0.814, and 0.786, well above the lower limit of 0.60 for research at exploratory stage [43], indicating adequate internal consistency among items within each factor (Table 2). All items load on the intended factor, with no cross-loadings in excess of 0.40. Based on the semantic meanings of core items under each factor, the three factors are labelled as natural identity, emotional affinity, and natural dependence, respectively. The three items that load on the natural identity factor are: "I am a part of nature, not separate from it", "I feel a sense of oneness with nature when I spend time in the wetland park" and "I feel a close connection with the environment". The second factor, emotional affinity, includes the following four items: "When I spend time in nature, I feel happy and satisfied", "When I spend time in nature, a feeling like security arises", "When I spend time in nature, I have a pleasant feeling of intimacy", and "When I spend time in nature, I feel a sense of awe for nature". The third factor, natural dependence consists of the following five items: "Connection to the environment in wetland is important to me", "I would feel that an important part of my life was missing if I could not enjoy nature from time to time", "I can derive spiritual sustenance from nature", "I need to be in the natural setting as much as possible" and "I spend a lot of time in natural settings". Therefore, tourists' connection to nature construct is examined as having three latent dimensions.

3.3 *Confirmatory Factor Analysis*

To test whether or not the dimensions identifies in the previous exploratory factor analysis could be verified across sub-samples, a confirmatory factor analysis (CFA) is conducted with the validation sample (n = 335). As reported in Table 3, the goodness-of-fit indices of measurement model (χ^2 = 99.418, *df* = 51, χ^2/df = 1.949, p < 0.001, RMSEA = 0.053, SRMR = 0.023, GFI = 0.955, NFI = 0.944, TLI = 0.972, CFI = 0.972) indicate a good fit between the CFA model and the validation sample. Standardized factor loadings for all items are above 0.50 on the constructs they are intended to measure [44], the t-values (ranging from 10.056 to 19.572) for all factor loadings are greater than 2.57 [45]. As shown in Table 3, all three factors reach the recommended level of construct reliability of 0.70 [42], with composite reliability values ranging from 0.738 to 0.800. The AVE values for the three constructs are 0.487, 0.564 and 0.448, which are considered reasonable for the newly developed

Table 2 Exploratory factor analysis results for tourists' connection to nature

Factor and item		Factor loading	Cronbach's α if deleted	Eigenvalue	Variance explained (%)	Cumulative variance explained (%)
Natural Identity (NI, α = 0.681)				4.995	41.626	41.626
NI1	I am a part of nature, not separate from it	**0.737**	**0.657**			
NI2	I feel a sense of oneness with nature when I spend time in the wetland park	**0.750**	**0.549**			
NI3	I feel a close connection with the environment	**0.737**	**0.539**			
Emotional Affinity (EA, α = 0.814)				1.225	10.205	51.831
EA1	When I spend time in nature, I feel happy and satisfied	**0.666**	**0.786**			
EA2	When I spend time in nature, a feeling like security arises	**0.782**	**0.738**			
EA3	When I spend time in nature, I have a pleasant feeling of intimacy	**0.816**	**0.742**			
EA4	When I spend time in nature, I feel a sense of awe for nature	**0.730**	**0.797**			
Natural Dependence (ND, α = 0.786)				1.030	8.587	60.418
ND1	Connection to the environment in wetland is important to me	**0.600**	**0.749**			

(continued)

Table 2 (continued)

Factor and item		Factor loading	Cronbach's α if deleted	Eigenvalue	Variance explained (%)	Cumulative variance explained (%)
ND2	I would feel that an important part of my life was missing if I could not enjoy nature from time to time	**0.718**	**0.744**			
ND3	I can derive spiritual sustenance from nature	**0.554**	**0.735**			
ND4	I need to be in the natural setting as much as possible	**0.727**	**0.741**			
ND5	I spend a lot of time in natural settings	**0.726**	**0.761**			
Total (**α = 0.866**)						

Note NI = natural identity; EA = emotional affinity; ND = natural dependence

scales [45]. Thus, the convergent validity of the latent dimensions is confirmed [44]. In addition, the discriminant validity of the scale is evaluated following Bagozzi and Phillips's suggestion [46]. Discriminant validity of the variables can be confirmed if the constrained model with the correlation coefficient of the two constructs set to "1" and the unconstrained model of each pair of constructs differ significantly on a χ^2 test. As can be seen from Table 4, the χ^2 test on each pair of constructs indicated significant differences at the 0.001 level. Further evidence of discriminant validity is provided by examining the confidence intervals of the paired correlations among the variables [47]. All confidence intervals do not include the value of 1 (Table 4), suggesting the discriminant validity of the measurement model is achieved.

To confirm whether the three-factor model is the more appropriate conceptualization of tourists' connection to nature, we conduct a CFA of a completing model with all items loading on one factor. As Table 5 shows, $\Delta\chi^2$ (3) = 186.628, $p < 0.001$, the one-factor model provides a significantly worse fit than the three-factor model. Thus, the three-factor model of tourists' connection to nature is supported.

Studies have indicated that connection to nature is a multi-dimensional construct [4, 20, 31]. In order to determine whether the three dimensions can be regarded as appropriate indicators of tourists' connection to nature, the second-order factor model is tested. Table 6 shows that the *t*-values (ranging from 9.101 to 13.139) of the

Table 3 Confirmatory factor analysis results for tourists' connection to nature

Dimension and item		SFL	*t*-Value	SMC	CR
Natural identity (NI)	NI1	0.612	11.008***	0.374	0.738
	NI2	0.770	14.425***	0.593	
	NI3	0.702	13.000***	0.493	
Emotional affinity (EA)	EA1	0.780	16.341***	0.609	0.837
	EA2	0.817	17.517***	0.668	
	EA3	0.880	19.572***	0.775	
	EA4	0.535	10.056***	0.286	
Natural dependence (ND)	ND1	0.793	16.396***	0.629	0.800
	ND2	0.682	13.238***	0.465	
	ND3	0.743	14.899***	0.552	
	ND4	0.633	12.082***	0.401	
	ND5	0.588	11.019***	0.346	

Model fit: $\chi^2 = 99.418$, $df = 51$, $\chi^2/df = 1.949$, $p < 0.001$, RMSEA = 0.053, SRMR = 0.023, GFI = 0.955, NFI = 0.944, TLI = 0.972, CFI = 0.972

Note SFL: standardized factor loading; SMC, squared multiple correlation; CR, composite reliability; RMSEA, root mean square error of approximation; SRMR, standardized root mean square residual; GFI, goodness of fit index; NFI, normed fit index; TLI, Tacker-Lewis index; CFI, comparative fit index

Table 4 Discriminant validity of the measurement models

Constructs	χ^2 Test		Confidence interval	
	$\Delta\chi^2$	Δdf	Lower	Upper
NI versus EA	89.309***	1	0.546	0.779
EA versus ND	111.548***	1	0.705	0.854
NI versus ND	82.367***	1	0.553	0.784

Note *** $p < 0.001$

Table 5 Model comparisons for dimensionality

Competing models	χ^2	*df*	*p* Value	GFI	NFI	TLI	CFI	SRMR	RMSEA
One-factor model	286.046	54	0.000	0.858	0.838	0.834	0.864	0.038	0.113
Three-factor model	99.418	51	0.000	0.955	0.944	0.963	0.972	0.023	0.053

Note The meanings of acronyms are the same as Table 3

Table 6 Second-order factor analysis of tourists' connection to nature

Second-order factor	First-order factor	SFL	*t*-Value
Tourists' connection to nature	NI	0.769	9.101***
	EA	0.873	12.874***
	ND	0.898	13.139***

three factor loadings are all significant at the 0.001 level. The measurement model of tourists' connection to nature has good fit indices. In all, results confirm that tourists' connection to nature is represented as an overarching concept consisting of natural identity, emotional affinity and natural dependence.

3.4 *Predictive Validity of Tourists' Connection to Nature Scale*

Besides assessing the construct validity of the scale of tourists' connection to nature, a test of nomological and predictive validity is further conducted by using structural equation modeling [42]. We combine the calibration sample and confirmatory sample ($n = 666$) to estimate a structural model in Chinese national wetland park with a path from tourists' connection to nature to pro-environmental behavior. The pro-environmental behavior is first subjected to an exploratory factor analysis using principal component method. The purpose of the EFA is to guarantee internal consistency of the construct in the context of Chinese national wetland park. The pro-environmental behavior is found as having two dimensions. The eight items load on two factors: general pro-environmental behavior and specific pro-environmental behavior. The Cronbach's alpha values for both two factors (0.856 and 0.741) are higher than 0.70, demonstrating adequate reliabilities.

The measurement model is tested using the whole sample. Following Anderson and Gerbing's two-step approach [48], CFA with the maximum likelihood estimation is employed to analyze how well the measured variables represent the constructs. The results show that the model fit the data reasonably well ($\chi^2 = 349.023$, $df = 163$, $\chi^2/df = 2.141$, $p < 0.001$, RMSEA = 0.041, GFI = 0.950, NFI = 0.938, TLI = 0.960, CFI = 0.966). The findings also indicate that tourists' connection to nature significantly influences pro-environmental behavior ($\beta = 0.940$, $p < 0.001$), explaining 88.4% of its variance. Therefore, the model confirm the nomological validity of the scale. The Standardized Root Mean Square Residual index (SRMR) is adopted to evaluate predictive validity [49]. As the SRMR value (0.039) is less than 0.08 [42], predictive validity is reached.

4 Conclusions and Implications

The study aims to provide insights into conceptualizing and measuring tourists' connection to nature in the context of Chinese national wetland parks. It also tries to examine the measurement predictive validity by investigating its implications on wetland tourists' pro-environmental behaviors. Following a rigorous process of multi-stage scale development, a three-dimension construct and 12-item scale of tourists' connection to nature is developed and empirically tested. The three dimensions are *Natural identity*, *Emotional affinity*, and *Natural dependence*. Comparison of the two competing models lends support to the proposed three-factor model, which reaches a good fit to the data. Furthermore, the findings of this study suggest that tourists' connection to nature significantly and directly influences their pro-environmental behaviors in the wetland parks. In conclusion, the study theoretically conceptualizes and empirically investigates the role of tourists' connection to nature in building tourists' pro-environmental behaviors. The findings reported in the study have both theoretical and practical implications.

4.1 Theoretical Implications

From a theoretical perspective, the scale of tourists' connection to nature provides a foundation for enhancing knowledge of tourist-nature relationships and extending theoretical understanding of tourists' connection to nature by empirically exploring its underlying dimensions. Unlike the independent self-construal in western culture, interdependent self-construal in Chinese culture provides a significant context to further develop the concept scope and complement the measurement scale of tourists' connection to nature [21, 23]. Guided by pre-identified constructs in relation to connection to nature in eco-psychology and environmental psychology, three underlying dimensions of tourists' connection to nature in the context of Chinese national wetland park are identified and empirically tested. Compared with existing scales, the present tourists' connection to nature scale has two advantages: (1) The three factors account for 60.418% of the total variance of tourists' connection to nature, well above the variance of the connectedness to nature scale (38%) [8] and the nature relatedness scale (34%) [31]; (2) Instead of integrating cognitive beliefs and emotional connections (e.g. nature relatedness scale), the study distinguishes the two aspects and extracted dimensions of natural identity and emotional affinity respectively. Among which, natural identity is commonly identified in prior research [4, 31], it captures the degree to which people identify with the natural community. Emotional affinity and its composing items have also been mentioned in the connection to nature literature [18, 32]. This factor reflects the level of tourists' emotional inclinations toward nature, such as feeling of happy, security, intimacy and a sense of awe for nature. (3) The dimension of natural dependence makes an important and original contribution to the literature. It is an important aspect of tourists' connection to nature, whereas rarely discussed in previous studies. The connotation of natural dependence is well-fitted in the principle of harmony and the doctrine of "the Unity of Heaven and Humanity" in traditional Chinese culture [24, 36]. As a key element of the connection to nature, natural dependence captures how Chinese people conceptualize their interdependent relationship with the natural world. It refers not only to people's belonging to a part of nature, but also an understanding of the importance of an integrated and united relationship between human and nature. In summary, the three-dimensional scale could be utilized as a research tool in future studies to assess how tourists connect to nature and belong to a part of the natural community.

Furthermore, the study provides a scale that can be used to investigate the effects of tourists' connection to nature on key tourist behavior variables. The results demonstrate that tourists' connection to nature predicts pro-environmental behavior, highlighting the predictive validity of the scale and the significance of fostering tourists' connection to nature in environmental planning. The study measures pro-environmental behavior as an outcome variable of tourists' connection to nature, and finds that tourists' connection to nature significantly and directly affects tourists' pro-environmental behaviors, suggesting that tourists' connection to nature could be a viable indicator in predicting pro-environmental behavior of wetland park tourists. The closer tourists' connecting to nature, the more likely they behave environmental friendly. This result aligns with the findings described in previous studies [19, 20, 38]. Theoretically, the study extends the investigation of tourists' connection to nature and pro-environmental behavior in the field of tourism and in the context of Chinese wetland parks, thus contributing to the existing literature on tourism environmental management.

4.2 Practical Implications

Several practical implications for destination managers and tourists can be derived from the findings. First, the three-dimensional scale of tourists' connection to nature can be used as a guidance for establishing strategies that encourage tourist's linkages to the natural environment. Tourists' connection to nature highlights the psychological significance of the relationships between human and nature, not just for well-being of nature, but for well-beings of tourists as well. Therefore, the implementation of mental health policies is urgently needed for the public mental health crisis after the Covid-19 pandemic. Practitioners can arrange targeted intervention by using natural elements to promote physical and psychological health, such as reshaping tourist's view of nature, taking nature-based treatments as an adjunctive therapy, and promoting nature-based activities. Second, the scale of tourists' connection to nature proves to be an effective tool for wetland managers to measure the extent to which tourist connects to nature. This measurement can be fulfilled by evaluating tourist's level of connection to nature before and after implementing a management program, which allows wetland managers to present a measurable justification for their future pro-environmental strategies. Moreover, the scale could be used to assess whether management interventions and environmental policies actually increase tourist's sense of connection to nature, such as attracting tourists to visit wetlands more frequently, increasing the amount of time tourists spent in wetlands, and providing environmental interpretation and education program on wetland and nature. In this respect, the development of the tourists' connection to nature scale provides meaningful values in tourism development and environmental management of wetland parks. In addition, the study demonstrates that tourists' connection to nature can effectively enhance tourist's pro-environmental behavior in a wetland park. It provides more insight on how tourists' connection to nature can contribute to goals of wetland environmental management. Tourists with strong connection to nature are more likely to act in a pro-environmental manner. Hence, wetland park managers should recognize tourists' connection to nature as a critical element in pursuing effective governance of wetlands. Managers should not only pay attention to improve the sense of connection to the park, but also formulate regulations to encourage pro-environmental behaviors and penalize behaviors that are harmful to the nature. For instance, managers could issue documentaries and advertisements of wetland parks and promote the recreational and restorative benefits of spending time in the wetland park through travel programs, media, and Internet.

4.3 Limitations and Directions for Future Research

This study contributes to the nature-based tourism and environmental psychology literature by offering a relatively reliable and valid measurement of tourists' connection to nature and providing further insights on the behavioral implications of the concept. However, the findings are subject to several limitations. First, the study adopted convenience sampling and conducted data collection at Xixi National Wetland Park. Further test in other tourism settings (e.g. natural conservation areas, forest areas, or wildlife sites) using a wide variety of sampling frames may afford greater generalizability of the results. Second, it should be noted that the study on pro-environmental behavior actually measures the self-reported behavioral intentions rather than actual behaviors. Future studies could include tourists' actual behavior through direct observation or tourist behavior recorders. Despite these limitations, this study provides directions for future research. Investigations on extended framework by incorporating other variables (e.g. environmental values, pro-environmental identity, and subjective norm) to examine causal relationships of environmental behavior could yield findings that have both theoretical values and practical significances.

Acknowledgements This study was funded by the Humanities and Social Sciences Research Project of Ministry of Education (No. 21YJC630089), the Basic Research Program of Guangzhou Science and Technology Bureau (No. SL2022A04J01735) and the Program of Philosophy and Social Science of Guangdong Province (No. GD24XGL010).

References

1. Dong, L., Bouey, J.: Public mental health crisis during COVID-19 pandemic. China. Emerg. Infect. Dis. **26**, 1616–1618 (2020)
2. Carvalho, P.M.D.M., Moreira, M.M., De Oliveira, M.N.A., Landim, J.M.M., Neto, M.L.R.: The psychiatric impact of the novel coronavirus outbreak. Psychiatr. Res. Neuroimag. **286**, 112902 (2020)
3. Li, W., Zhang, H., Zhang, C., Luo, J., Wang, H., Wu, H., Zhu, Y., Cui, H., Wang, J., Li, H., Zhu, Z., Xu, Y., Li, C.: The prevalence of psychological status during the COVID-19 epidemic in china: a systemic review and meta-analysis. Front. Psychol. **12**, 614964 (2021)
4. Clayton, S.: Environmental Identity: a Conceptual and Operational Definition. In: Clayton. S., Opotow, S. (eds.) Identity and the natural environment. pp. 45–65. Cambridge, MIT Press (2003)
5. Huynh, T., Torquati, J.C.: Examining connection to nature and mindfulness at promoting psychological well-being. J. Environ. Psychol. **66**, 101370 (2019)
6. Barlett, P.F.: Reason and reenchantment in cultural change: sustainability in higher education. Curr. Anthropol. **49**, 1077–1098 (2008)
7. Mayer, F.S., Frantz, C.M., Bruehlman-Senecal, E., Dolliver, K.: Why is nature beneficial? Environ. Behav. **41**, 607–643 (2009)
8. Mayer, F.S., Frantz, M.: The connectedness to nature scale: a measure of individuals' feeling in community with nature. J. Environ. Psychol. **24**(4), 503–515 (2004)
9. McEwan, K., Ferguson, F.J., Richardson, M., Cameron, R.: The good things in urban nature: a thematic framework for optimising urban planning for nature connectedness. Landscape Urban Plan. **194**, 103687 (2020)
10. Boiral, O., Heras-Saizarbitoria, I., Brotherton, M.C.: Nature connectedness and environmental management in natural resources companies: an exploratory study. J. Clean. Prod. **206**, 227–237 (2019)
11. Bamberg, S., Moser, G.: Twenty years after Hines, Hungerford, and Tomera: a new meta-analysis of psycho-social determinants of pro-environmental behavior. J. Environ. Psychol. **27**, 14–25 (2007)
12. Winter, D.D.: Some big ideas for some big problems. Am. Psychol. **55**, 516–522 (2000)
13. Tacey, D.J.: Re-enchantment: the New Australian spirituality. Harper Collins, Sydney (2000)
14. Leopold, A.: A sand County Almanac: with essays on conservation from round river. Ballantine Books, New York (1949)
15. Kellert, S.R., Wilson, E.O.: The biophilia hypothesis. Island Press, Washington, D.C. (1993)
16. Roszak, T.: Where psyche meet Gaia. In: Roszak, T., Gomes, M.E., Kanner A.D. (eds.) Ecopsychology: restoring the earth, healing the mind. Sierra Club Books, San Francisco (1995)
17. Collado, S., Evans, G.W.: Outcome expectancy: a key factor to understanding childhood exposure to nature and children's pro-environmental behavior. J. Environ. Psychol. **61**, 30–36 (2019)
18. Kals, E., Schumacher, D., Montada, L.: Emotional affinity toward nature as a motivational basis to protect nature. Environ. Behav. **31**(2), 178–202 (1999)
19. Perkins, H.E.: Measuring love and care for nature. J. Environ. Psychol. **30**(4), 455–463 (2010)
20. Tam, K.P.: Concepts and measures related to connection to nature: similarities and differences. J. Environ. Psychol. **34**, 64–78 (2013)
21. Schultz, P.W., Shriver, C., Tabanico, J., Khazian, A.: Implicit connections with nature. J. Environ. Psychol. **24**(1), 31–42 (2004)
22. Clayton, S.: Environment and Identity. In: Clayton, S. (ed.) The Oxford handbook of environmental and conservation psychology, pp. 164–180. Oxford University Press, New York (2012)
23. Singelis, T.M.: The measurement of independent and interdependent self-construals. Pers Soc Psychol B. **20**(5), 580–591 (1994)
24. Buckley, R., Cater, C., Zhong, L., Chen, T.: Shengtai Luyou: cross-cultural comparison in ecotourism. Ann Tourism Res. **35**(4), 945–968 (2008)

25. Qian, X.: Harmony: each in its due place. J. Tsinghua Univ. (Philos. Soc. Sci.) **5**, 3–8 (2001)
26. Li, F.M.S.: Culture as a major determinant in tourism development of China. Curr. Issues Tour. **11**(6), 492–513 (2008)
27. Xu, H., Cui, Q., Sofield, T., Li, F.M.S.: Attaining harmony: understanding the relationship between ecotourism and protected areas in China. J. Sustain. Tour. **22**(8), 1131–1150 (2014)
28. Bragg, E.A.: Towards ecological self: deep ecology meets constructionist self-theory. J. Environ. Psychol. **16**(2), 93–108 (1996)
29. Perrin, J.L., Benassi, V.A.: The connectedness to nature scale: a measure of emotional connection to nature? J. Environ. Psychol. **29**(4), 434–440 (2009)
30. Schultz, P.W.: The structure of environmental concern: concern for self, other people, and the biosphere. J. Environ. Psychol. **21**(4), 327–339 (2001)
31. Nisbet, E.K., Zelenski, J.M., Murphy, S.A.: The nature relatedness scale: linking individuals' connection with nature to environmental concern and behavior. Environ. Behav. **41**(5), 715–740 (2009)
32. Davis, J.L., Green, J.D., Reed, A.: Interdependence with the environment: commitment, interconnectedness, and environmental behavior. J. Environ. Psychol. **29**(2), 173–180 (2009)
33. Dutcher, D.D., Finley, J.C., Luloff, A.E., Johnson, J.B.: Connectivity with nature as a measure of environmental values. Environ. Behav. **39**(4), 474–493 (2007)
34. Davis, J.L., Le, B., Coy, A.E.: Building a model of commitment to the natural environment to predict ecological behavior and willingness to sacrifice. J. Environ. Psychol. **31**(3), 257–265 (2011)
35. Tang, Y., Li, Z., McLean, G.F.: Man and nature: the chinese tradition and the future. University Press of America, Lanham (1989)
36. Tu, W.: The Ecological Turn in New Confucian Humanism: Implications for China and the World. Daedalus: J AAAS. 130(4), 243–264 (2001)
37. Sofield, T., Li, F.M.S.: China: ecotourism and cultural tourism: harmony or dissonance? In: Higham, J. (ed.) Critical issues in ecotourism: confronting the challenges, pp. 368–385. Elsevier Science & Butterworth Heinemann, London (2007)
38. Hehir, C., Scarles, C., Wyles, K.J., Kantenbacher, J.: Last chance for wildlife: making tourism count for conservation. J. Sustain. Tour. (2022). https://doi.org/10.1080/09669582.2022.2049804
39. Stewart, A.M., Craig, J.L.: Predicting pro-environmental attitudes and behaviors: a model and a test. J Environ Syst. 28(4), 293–317 (2000/2001)
40. Kim, J.-H., Ritchie, J.R.B., McCormick, B.: Development of a scale to measure memorable tourism experiences. J. Travel Res. **51**(1), 12–25 (2012)
41. Cheng, T.M., Wu, H.C., Huang, L.M.: The influence of place attachment on the relationship between destination attractiveness and environmentally responsible behavior for Island tourism in Penghu. Taiwan. J. Sustain. Tour. **21**(8), 1166–1187 (2013)
42. Hair, J.F., Black, W.C., Babin, B.J., Anderson, R.E.: Multivariate data analysis (7th ed.). Pearson Prentice Hall, Upper Saddle River (2009)
43. Nunnally, J.C., Bernstein, I.H.: Psychometric theory, 3rd edn. McGraw-Hill, New York (1994)
44. Fornell, C., Larcker, D.F.: Evaluating structural equation models with unobservable variables and measurement error. J. Marketing Res. **18**, 39–50 (1981)
45. Netemeyer, R.G., Bearden, W., Sharma, S.: Scaling procedures: issues and applications. Sage, Thousand Oaks (2003)
46. Bagozzi, R.P., Phillips, L.W.: Representing and testing organizational theories: a holistic construal. Admin. Sci. Quart. **27**(3), 459–489 (1982)
47. Torkzadeh, G., Koufteros, X., Pflughoeft, K.: Confirmatory analysis of computer self-efficacy. Struct. Equ. Modeling **10**(2), 263–275 (2003)
48. Anderson, J.C., Gerbing, D.W.: Structural equation modeling in practice: a review and recommended two-step approach. Psychol. Bull. **103**(3), 411–423 (1988)
49. Dedeoglu, B.B., Taheri, B., Okumus, F., Gannon, M.: Understanding the importance that consumers attach to social media sharing (ISMS): scale development and validation. Tourism Manage. **76**, 103954 (2020)

Analysis of Factors Influencing the Academic Conference Participation

Jorge Marques and Makhabbat Ramazanova

Abstract This article aims to explore the main factors that influence academics' decision to participate in academic and scientific events. A quantitative methodology is developed by applying online questionnaires to academics who regularly participate in scientific events. Results show that the factors that tend to influence the decision to participate in scientific events can be grouped into four types: organizational context, academic opportunities, opportunities for leisure, recreation, and culture, and event accessibility. The originality of this work lies in the updated identification and consideration of recent events related to the COVID-19 pandemic, as well as the main factors influencing the decision of choice related to the scientific events in which academics participate.

Keywords Conference tourism · Academic and scientific events · Decision to attend · Factors

1 Introduction

For scholars, it is essential to participate in conferences, as it represents the possibility to share new knowledge, to keep updated in their fields of work and research, and to improve networking, contributing thus to building up academic identity and reputation [1, 2]. However, the COVID-19 pandemic has placed higher restrictions on scientific and academic events and, thus, conference attendance [3]. Physical (face-to-face) events were cancelled, postponed, and often replaced by online events. The virtualization of events has profoundly changed how academics meet and debate the most

J. Marques · M. Ramazanova
REMIT—Research On Economics, Management and Information Technologies, University Portucalense Infante D. Henrique, Porto, Portugal
e-mail: ramazanova@upt.pt

J. Marques (✉)
CEGOT—Centre of Studies in Geography and Spatial Planning, University of Coimbra, Coimbra, Portugal
e-mail: jorgem@upt.pt

Y. Luo et al. (eds.), *Tourist Experience and Perception in the Pandemic and Post-Pandemic Era*, https://doi.org/10.1007/978-981-96-1820-0_6

diverse topics. However, the literature review shows that face-to-face contact, or co-presence, is essential for networking and has successful implications in the workplace [4]. Particularly important for academics and their performance criteria, networking and collaboration opportunities at academic conferences are highly relevant [2, 4]. As such, it is essential to understand better the main factors influencing participation in this event. In this context, this study explores the main factors that influence academics' decision to participate in academic and scientific events. Following this brief introduction section, the literature review is presented on the topic. The methodology section demonstrates the sampling, data collection and analysis processes. Section four is focused on results and discussion, followed by concluding remarks and limitations of the study.

2 Literature Review

Researchers have long worked to understand better the factors that influence the organizer's decision to choose the location for conferences or conventions while also trying to understand which factors, from the participants' perspective, influence the process of choosing the events in which they participate. One of the first research works on the factors that influence convention site selection was presented by Fortin and Ritchie (1977) [5]. However, the results were limited, mainly showing that corporations and associations differ significantly in their motivations and processes for selecting a host convention site. On the other hand, Oppermann and Chon (1997) [6] developed a conceptual framework for convention participation decision-making by incorporating empirical findings from previous convention tourism literature. The authors pointed out four factors entailed in the decision-making process of an individual with predispositions toward attending a convention: (i) personal or business factors, (ii) association or conference factors, (iii) location factors, and (iv) intervening factors. Crouch and Ritchie's (1998) [7] studies on convention site selection identified in the literature eight main categories of factors: accessibility, local support, extra-conference opportunities, accommodation facilities, meeting facilities, information, site environment, and other criteria. From this point, Crouch and Ritchie (1998) [7] developed their conceptual model, which was based on three main steps: (i) convention preplanning, (ii) site selection analysis and recommendations, and (iii) site selection decision.

Based on the literature review, Crouch and Louviere (2004) [8] identified 20 different convention site attributes affecting the selection of a host destination for a convention. Results suggested that although meeting facility characteristics is significant, an attractive host site must also offer strengths across various factors to succeed in a competitive environment. Destinations must create unique combinations of attributes to develop competitive, solid positions. They must do so using a deep knowledge of the factors that event organizers value most in their site choice decisions. Yoo and Chon (2008) [9] developed a five-factor scale for convention participation decision-making: (i) destination stimuli, (ii) professional and social networking opportunities, (iii) educational opportunities, (iv) safety and health situation, and (v) travelability. More recently, after a review of academic literature and industry reports, Cassar, Whitfield and Chapman (2020) [10] identified the motivators and inhibitors influencing association conference attendance according to three specific groups of factors: (i) Destination-related factors, namely, conference facilities, accessibility, climate, destination's image and positioning, and safety; (ii) Delegate-related factors, such as individual's involvement in the hosting organization, access to external funding for attendance, health conditions of the attendants, past experiences, travel ability, and personal budget; (iii) Conference-related factors, including the potential for educational improvement, career enhancement, the prospect of developing new skills, networking and socializing opportunities, and the provision of leisure-related activities.

3 Methodology

The study explored the factors influencing academics' decisions to participate in academic and scientific events in the post-pandemic era. A quantitative methodology was adopted, involving the distribution of online questionnaires. The design and distribution of the survey questionnaire were informed by previous studies on similar topics, integrating motivational elements identified in the literature [2, 4, 7, 8, 11–14].

The survey consisted of 16 items measured on a 5-point Likert scale, with responses ranging from "Very Important" (5) to "Not at all Important" (1). The first part of the questionnaire collected socio-demographic information (gender, age, work status, education level, position, and work institution); the second part focused on elements influencing the decision to attend academic events.

Participants were selected using a convenience sampling approach, targeting academics previously attending scientific events. The survey was distributed via email and posted in selected Facebook academic groups during April 2020. Two hundred thirty-five questionnaires were collected, of which 226 were valid and included in the analysis.

Data was analysed using the Statistical Package for Social Sciences (SPSS, version 27). Descriptive statistics were calculated for socio-demographic variables and factors influencing conference participation. An exploratory factor analysis (EFA) was conducted to identify underlying factors influencing the decision to participate in academic events. The EFA utilised principal component analysis (PCA) with varimax rotation. The Kaiser–Meyer–Olkin (KMO) measure and Bartlett's test of sphericity were used to assess the suitability of the data for factor analysis. The KMO value was 0.836, and Bartlett's test was significant ($p < 0.001$), indicating that the data were appropriate for factor analysis.

Four factors with eigenvalues greater than 1 were extracted, explaining 65% of the total variance. The factors were labelled based on the attributes that loaded significantly on each factor:

1. Conference-related Factors: These include items such as the quality of accommodation, food, venue, event support services, and attending the event at the same hotel.
2. Personal-related Factors: Encompassed opportunities to present and publish academic work, quality of the event's academic program, and networking opportunities.
3. Destination-related Factors: Comprised opportunities to learn about local culture, recreational and leisure activities, and the destination's reputation.
4. Accessibility-related Factors: Included travel distance and costs associated with participation.

Reliability analysis used Cronbach's alpha, with values above 0.75 for the first three factors, indicating good internal consistency. The fourth factor, while having a lower reliability (Cronbach's alpha = 0.54), was retained due to its reasonable factor loading and eigenvalue greater than 1.

This methodology provided a comprehensive approach to understanding the multifaceted decision-making process of academics regarding conference participation, considering both personal and contextual factors.

4 Results and Analysis

4.1 Socio-Demographic Characteristic of Respondents

Of the sample, 58.4% were female, and 40.7% were male, while 0.9% preferred target no answer. Most respondents (85%) are between 31 and 60 years old, including academics at different stages of their careers. They work mainly in public (55,8%) and private (35%) universities, with a small percentage being employed by research institutes (5,8%), with 86,3% working full-time. Most academics hold a PhD (54,9%), and 19,5% have post-doctoral studies. Regarding the position, most of the respondents are teaching staff (71%), followed by administrative staff, researchers, graduate students, PhD students and others (Table 1).

As for country of residence, 48% of respondents are from Portugal, and 52% are from other countries. In addition to Europe, which is more represented, respondents from the Americas, Africa, the Middle East, Asia, and Oceania were also collected. It is valuable to note the average annual number of trips usually taken to participate in academic and scientific events (with an overnight stay of at least one night), which on average is 3.62, being a minimum 1 trip and a maximum of 26.

Various sources usually support the cost associated with participation in the event. Figure 1 shows the main ways of financing academic conference attendance.

Although there are various sources of financing for participation in academic and scientific conferences, the one that was most often indicated was paid from own pocket, followed by financing supported research centres. It will be essential to mention that each respondent could select more than one form, and, as such, it

Table 1 Sample profile

Category	Description	Frequency	%
Gender	Female	132	58
	Male	92	41
	Rather not answer	2	1
Age	20–30	14	6
	31–40	66	29
	41–50	73	32
	51–60	53	23
	61+	20	9
Work status	Full-time	195	86
	Part-time	31	14
Education level	Graduation	8	4
	Postgraduate studies	11	5
	Master's degree	36	16
	Doctorate	124	55
	Post-Doctorate	44	19
	Others	3	1
Position	Teaching staff	161	71
	Administrative staff	4	2
	Researcher	26	12
	Graduate Student	4	2
	PhD/Post-Doctoral Student	23	10
	Other	8	4
Work institution	Public universities	126	56
	Private universities	79	35
	Research institutes	13	6
	Others	8	4

was verified that the participants had several associated forms of payment in many cases. This means that conference attendants support participation often in their own pockets; sometimes, they go through research centres' funding or from projects that they are involved with. In any case, the high number of participants who indicated they were paying from their pocket looks relevant to us.

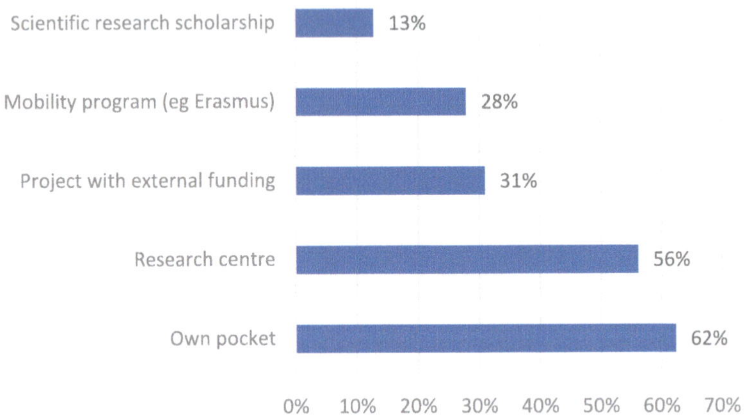

Fig. 1 Main sources of financing for participation in academic conferences

4.2 Factor Analysis

The previous studies illustrated that motivators and inhibitors influencing association conference attendance are diverse. For this research, 16 items related to the academic events participation motivation were assessed, and an exploratory factor analysis using a varimax rotation with principal components analysis (PCA) extraction was used to group them into a smaller number of factors explaining participants' decisions to participate in the events. Means for these items ranged from 4.67 (highest) to 2.72 (lowest) and are presented in Table 2. Quality of the event's academic and scientific program, networking opportunities and interpersonal relationships, opportunities to present and public scientific projects are assessed as very important by more than 60% of the respondents with a mean of about 4.50.

Bartlett's test of sphericity and Kaiser-Meyer-Oklin Measure of Sampling Adequacy are used to assess the feasibility of applied factor analysis. The Kaiser–Meyer–Olkin Measure of Sampling Adequacy is a statistic indicating the proportion of variance in the variables underlying factors might cause. High values indicate a significant correlation between original variables, meaning a factor analysis may be helpful. The results of the KMO and Bartlett's Test in our study (0.836 and $p < 0.001$) revealed that the factor analysis was appropriate [15] (Table 3).

Principal Component Analysis revealed the presence of 4 factors with eigenvalues greater than 1.0, explaining 65% of the cumulative variance (Table 4). The rotation varimax was used to exhibit the loadings, and variables that loaded at a correlation level of 0.5 were included in the present analysis [16].

As for the reliability of the factors, Cronbach's alpha, the most used reliability measure, was then applied to test the reliability of the extracted factors. The Cronbach's alpha values are higher than 0.75 for factors 1, 2, and 3, which exceeds the

Table 2 Factors influencing the decision to participate in an academic event

Survey items	1 (%)	2 (%)	3 (%)	4 (%)	5 (%)	Mean
Quality of the event's academic/scientific program	0	0.9	2.7	25.2	71.2	**4.67**
Networking opportunities and interpersonal relationships	0	1.3	6.6	30.5	61.5	**4.52**
Opportunities to present academic/scientific work	1.3	1.8	5.3	28.8	62.8	**4,50**
Opportunities for publishing academic/scientific papers	2.2	3.1	8	20.8	65.9	**4.45**
Costs associated with participation in the event (registration, accommodation, transport)	0.9	3.1	15.9	36.3	43.8	**4.19**
Perceived safety of the destination	0	11.9	11.5	38.5	38.1	**4.03**
Opportunity to learn about local culture and heritage	0.9	7.5	16.8	47.8	27	**3.92**
Event support services (event information, secretariat, logistical support, technical support...)	5.8	15	22.6	36.7	19.9	**3.50**
Destination reputation	8	14.2	28.8	34.1	15	**3.34**
Distance from where the event takes place (travel distance)	12.4	16.4	27.4	23.5	20.4	**3.23**
Quality of the venue where the event takes place	10.6	19	27.4	31.4	11.5	**3.14**
Opportunity to develop recreational and leisure activities	8.8	22.6	31	27	10.6	**3.08**
Quality of accommodation	11.1	18.1	32.3	30.1	8.4	**3.07**
Quality of food	14.2	19.9	34.1	23.9	8	**2.92**
Distance between the venue and the nearest airport	18.6	27	20	22	12.4	**2.83**
Attending the event at the same hotel where you are staying	26.1	21,7	18.1	22.1	11.9	**2.72**

1, Not important at all; 2, Not important; 3, Netral; 4, Important; 5, Very important

Table 3 KMO and Bartlett's test

Kaiser–Meyer–Olkin measure of sampling adequacy		0.836
Bartlett's test of sphericity	Approx. Chi-Square	1524.350
	Df	120
	Sig	0.001

recommended level of 0.7 [17] and illustrates acceptable levels of internal consistency of these factors. To maintain the exploratory power of the extracted factors, factor 4 was retained for this analysis also, despite its reliability of 0.54, explaining that due to its reasonable factor loading and the eigenvalue higher than 1 [18].

As shown in Table 5, the predominant factor is factor 1, labelled "Conference-related factors". The items included in this factor are related to the quality of accommodation, food, venue, location of the hotel, distance from the airport and event

Table 4 Total variance explained

Initial Eigen values			
Component	Total	% of Variance	Cumulative %
1	5,145	32,156	32,156
2	2,388	14,925	47,081
3	1,513	9,459	56,540
4	1,363	8,519	65,059
Extraction sums of squared loadings			
Component	Total	% of Variance	Cumulative %
1	5,145	32,156	32,156
2	2,388	14,925	47,081
3	1,513	9,459	56,540
4	1,363	8,519	65,059
Rotation sums of squared loadings			
Component	Total	% of Variance	Cumulative %
1	3,833	23,959	23,959
2	2,500	15,622	39,581
3	2,311	14,444	54,025
4	1,766	11,034	65,059

Extraction method: principal component analysis

support services such as information, secretariat, logistical support, and technical support. The second factor is titled "Personal related factors", which assesses several academic advantages the participants see in such events. This factor covers opportunities to present academic and scientific work, publish research results, network opportunities and interpersonal relationships, and improve the event program's quality. The third factor was "Destination factors", including items related to the opportunity to learn about local culture and heritage, to develop recreational and leisure activities and destination reputation. The fourth factor was titled "Accessibility-related factors", incorporating a distance from where the event takes place (travel distance) and costs associated with participation in the event (registration, accommodation, transport). These results align with a literature review, for most two authors refer to these factors as the most decisive in the decision-making process associated with participation in conferences [2, 4, 7, 8, 11–14]. More specifically, in the case of academic conferences, the results are similar and support this review.

Table 5 Factor analysis

Factor name	Attributes	Loadings	Eigenvalue	% of variance	% of cumulative variance	Cronbach alpha
F1: Conference related factors	Quality of accommodation	0.872	5.145 0	32.20	32.20	0.870
	Quality of food	0.868				
	Quality of the venue where the event takes place	0.850				
	Attending the event at the same hotel where you are staying	0.701				
	Event support services (information, secretariat, logistical, technical support…)	0.686				
F2: Personal related factors	Opportunities to present academic/ scientific work	0.823	2.388	14.92	47.08	0.753
	Opportunities for publishing academic/ scientific papers	0.822				
	Quality of the event's academic/ scientific program	0.664				
	Networking opportunities and interpersonal relationships	0.653				
F3: Destination related factors	Opportunity to learn about local culture and heritage	0.820	1.513	9.50	56.54	0.763
	Opportunity to develop recreational and leisure activities	0.814				
	Destination reputation	0.686				

(continued)

Table 5 (continued)

Factor name	Attributes	Loadings	Eigenvalue	% of variance	% of cumulative variance	Cronbach alpha
F4: Accessibility related factors	Distance from where the event takes place (travel distance)	0.847	1.363	8.52	65.05	0.544
	Costs associated with participation in the event (registration, accommodation, transport)	0.643				

Absolute loadings less than 0.5 have been omitted
Extraction method: principal component analysis. rotation method: varimax with Kaiser Normalization

5 Conclusions

The current work proves that participation in academic events can be motivated by diverse reasons. The conference-related factor demonstrates the importance of quality of accommodation, food, the venue where the event takes place, event support services and convenient location in terms of holding the event at the same hotel where staying. The second factor illustrated high personal motivation in attending conferences, such as opportunities to present scientific work and publications, networking opportunities, and interpersonal relationships.

In turn, destination-related factors are essential, such as the opportunity to learn about local culture and heritage and to develop recreational and leisure activities during a stay in a specific destination. Combining a work visit with leisure and recreation activities is motivational for academics. The fourth factor underlines that accessibility influences the decision of the participants in terms of distance from where the event takes place and costs associated with participation in the event (registration, accommodation, transport).

Building on this study and the existing literature, it becomes evident that the topic of motivation to attend academic events is a rich area for further exploration. This study serves as a stepping stone, but more research is needed, particularly to gauge the medium and long-term impacts of the COVID-19 pandemic on event participation.

Acknowledgements This work was supported by the FCT Fundação para a Ciencia e a Tecnologia, I.P. [Project UIDB/05105/2020].

References

1. Edelheim, J., Thomas, K., Åberg, K., Phi, G.: What do conferences do? What is academics' intangible return on investment (ROI) from attending an academic tourism conference? J. Teach. Travel Tour. **18**(1), 94–107 (2017). https://doi.org/10.1080/15313220.2017.1407517
2. Mair, J., Lockstone-Binney, L., & Whitelaw, P.: The motives and barriers of association conference attendance: Evidence from an Australasian tourism and hospitality academic conference. J. Hospital. Tourism Manage. 34, 58–65 (2018). ISSN 1447–6770, https://doi.org/10.1016/j.jhtm.2017.11.004
3. Gössling, S., Scott, D., Hall, M.: Pandemics, tourism and global change: a rapid assessment of COVID-19. J. Sustain. Tour. **29**(1), 1–20 (2020). https://doi.org/10.1080/09669582.2020.1758708
4. Severt, D., Wang, Y., Chen, P.-J., Breiter, D.: Examining the motivation, perceived performance, and behavioural intentions of convention attendees: evidence from a regional conference. Tour. Manage. **28**, 399–408 (2006). https://doi.org/10.1016/j.tourman.2006.04.003
5. Fortin, P.A., Brent Ritchie, J.R.: An empirical study of association decision processes in convention site selection. J. Travel Res. **15**(4), 13–20 (1977). https://doi.org/10.1177/004728757701500403
6. Oppermann, M., Chon, K.: Convention participation decision-making process. Ann. Tour. Res. **24**(1), 178–191 (1997). https://doi.org/10.1016/S0160-7383(96)00049-7
7. Crouch, G., Ritchie, J.: Convention site selection research: a review, conceptual model and propositional framework. J. Conven. Exhibit. Manage. **1**(1), 49–69 (1998). https://doi.org/10.1300/J143v01n01_05
8. Crouch, G., Louviere, J.: The determinants of convention site selection: a logistic choice model from experimental data. J. Travel Res. **43**(2), 118–130 (2004). https://doi.org/10.1177/0047287504268233
9. Yoo, J.J.-E., Chon, K.: Factors affecting convention participation decision-making: developing a measurement scale. J. Travel Res. **47**(1), 113–122 (2008). https://doi.org/10.1177/0047287507312421
10. Cassar, J., Whitfield, J., & Chapman, A.: Contemporary factors influencing association conference attendance. J. Convent. Event Tourism 21, 57–90 (2020). https://doi.org/10.1080/15470148.2020.1719948
11. Jago, L.K., Deery, M.: Relationships and factors influencing convention decision-making. J. Conv. Event Tour. **7**(1), 23–41 (2005)
12. Mair, J., Thompson, K.: The UK association conference attendance decision-making process. Tour. Manage. **30**(3), 400–409 (2009). https://doi.org/10.1016/j.tourman.2008.08.002
13. Rittichainuwat, B.N., Beck, J.A., Lalopa, J.: Understanding motivations, inhibitors, and facilitators of association members in attending international conferences. J. Conven. Exhibit. Manage. **3**(3), 45–62 (2001). https://doi.org/10.1300/J143v03n03_04
14. Yoo, J., Zhao, X.: Revisiting determinants of convention participation decision making. J. Travel Tour. Mark. **27**(2), 179–192 (2010). https://doi.org/10.1080/10548401003590369
15. Napitupulu, D., Kadar, J.A., Jati, R.K.: Validity testing of technology acceptance model based on factor analysis approach. Indones. J. Electr. Eng. Comput. Sci. **5**(3), 697–704 (2017)
16. Hair, J., Anderson, R.E., Tatham, R.L., Black, W.C.: Multivariate data analysis, 4th edn. Prentice-Hall Inc., New Jersey (1995)
17. Nunnally, J.C.: Psychometric theory, 2nd edn. McGraw-Hill, New York (1978)
18. Kaiser, H.F.: A computational starting point for Rao's canonical factor analysis: Implications for computerized procedures. Educ. Psychol. Measur. **34**(3), 691–692 (1974). https://doi.org/10.1177/001316447403400322
19. Wu, L., Waber, B., Aral, S., Brynjolfsson, E., Pentland, A.: Mining face-to-face interaction networks using sociometric badges: predicting productivity in an IT configuration task. ICIS 2008 Proceedings. 127 (2008). https://aisel.aisnet.org/icis2008/127

The Mechanism of Decision-Making Behavior of Ski Tourists

Jiaqi Chen, Xi Li, and Chunling Liu

Abstract Role identity, place attachment, and perceived value are social psychology concepts that have become research hotspots as place attachment and perceived value have been introduced into tourism research in the past two years. On this basis, this paper further expands the research perspective and attempts to introduce role identity into the field of tourism. The main idea is to incorporate place attachment and perceived value as common variables in ski tourist decision-making behavior. The paper also constructs a structural model with perceived value as the antecedent variable, role identity and place attachment as the mediator variables, and decision-making behavior as the outcome variable. It takes the Winter Olympics hosting place, Chongli, Zhangjiakou, as a case study, using the SEM method to test the mechanism of action between the variables. The results of the survey show that ski tourists' role identity and place attachment completely mediate the influence of tourism's perceived value on decision-making behavior. Moreover, the self-identity dimension of role identity and the place identity and social bond dimensions of place attachment have a significant direct influence on decision-making behavior. The results of this study revise the conclusion that perceived value has a direct effect on decision-making behavior.

Keywords Role identification · Perceived value · Place attachment · Decision-making behavior · Ski tourists

J. Chen · C. Liu
Tourism College, Guangzhou City Polytechnic, Guangzhou 511370, China
e-mail: 1806388359@qq.com

X. Li (✉)
Faculty of International Tourism and Management, City University of Macau, Taipa, Macau S.A.R., China
e-mail: xli@cityu.edu.mo

Y. Luo et al. (eds.), *Tourist Experience and Perception in the Pandemic and Post-Pandemic Era*, https://doi.org/10.1007/978-981-96-1820-0_7

1 Introduction

This study is based on China's successful bid to host the 2022 Winter Olympic Committee, the government's strategic guidance, and the market's enthusiastic response to the development of the domestic ski industry to bring vigorous vitality, skiing from competitive sports to mass leisure sports rapid transformation, this phenomenon has been the attention of academics and has become a hot spot for research. Stimulated by the Olympic event, in recent years, the domestic ski group scale significantly expanded, the number of ski resorts surge, but at the same time, China's skiing scenic spots generally exist in the product level is low, homogenization is obvious, the supporting facilities and equipment, management and service and cannot effectively meet the needs of ski tourists, so that their identity of the skiing destination is low, etc. become a constraint on the development of the outstanding problems. As a result, understanding the consumption behavior of ski tourists, innovating ski products, and providing products and services that meet their psychological needs are necessary for the sustainable development of ski destinations. Role identity, perceived value, and place attachment are important indicators for predicting tourists' demand-influenced decision-making behavior. Place attachment has an important impact on tourists' image identity and post-trip loyalty, which has become a hot topic of discussion in tourism and leisure [1]. Role identity is the basis of identity theory and role line analysis, used in corporate employee traits [2] college student differences [3], volunteer role behaviour [4] identification research, tourism and sports and leisure is an emerging field less involved. Perceived value, and decision-making behavior, on the other hand, is one of the most researched theses in the field of tourism in recent years.

2 Literature Review

2.1 Perceived Value and Consumer Behavior

Based on the exchange theory to explain the customer value perception, reflects the customer in the process of exchange of goods or services for money, time, and other costs of payment and gain the degree of importance of the benefits. Initial research defined the connotation of customer-perceived value based on the trade-off between customer-perceived gains and losses, emphasizing the individuality and subjective feeling of customer value [5]. Further research has shown that customer perceived value is a dynamic change, customer perception of the purchase or use of products or services, functions, attributes, and utility, to achieve the expected results of a comprehensive evaluation, to determine whether it meets the expected psychological needs and its results of the assessment [6]. Further discussion based on existing research, the first interpretation of perceived value from the perspective of monetary prices, is the customer perception of product utility quality benefits and consumption of economic costs of the trade-off between the gains and losses, and that the perceived value that is the ratio between the profits and losses [7]. Kotler elaborates on customer value from the perspective of the value of alienation, which solves the difficult problem of the inability of Zeithaml to measure and estimate the customer's benefits and costs, and that the total cost of the customer does not only refer to the monetary aspect, but also includes time, mental, and physical costs, and the total cost of the customer. monetary aspects, but also non-monetary cost inputs in terms of time, and mental and physical effort [8]. The attempt to rationalize the assessment of gains and losses in the actual consumption process appears to be more complex, Sheath divided the dimensions of perceived value assessment into six aspects: social value, emotional value, functional value, cognitive value, and conditional value [9], Sweeney defined the dimensions as affective value, social value, quality, and price [10], the Yonggui suggested that perceived value should be comprehensively judged in four dimensions: emotional value, social value, functional value, and cost [11].

Since the mid-1990s, the study of perceived value has been gradually expanded to the field of tourism, Morrison defined the connotation of tourists' perceived value earlier, arguing that tourists' perceived value is a comparative analysis of the products or services purchased by tourists in the tourism process in terms of the actual benefits obtained and the cost of inputs, which leads to the formation of a psychological evaluation in the hearts of tourists[12]. Gallarza argues that tourists constantly make value judgments about the emotional and cognitive image of the destination during the tourism process, which in turn is an important medium for generating behavioral intentions [13]. This view implies that value judgments formed by customers' actual perception of a tourist destination can influence customers' future behavioral intentions and become an important antecedent that affects the willingness to revisit promote and recommend the destination to others.

The measurement of perceived value focuses on the consideration of both psychological and economic factors, and the measurement dimension is expanded from a single dimension to a multi-dimensional one. The perceived value scale for ski tourists in the study is mainly derived from Pan [14] and the research results of domestic scholars such as Zhang Tao and Bao Jigang [15], including four dimensions social value, emotional value, knowledge value, and functional value.

2.2 *The Influence of Place Attachment on Tourism Behavior*

As the research on place attachment in tourism behavior continues to expand and deepen, its influence on tourism behavior has attracted more and more attention. Tourist destinations with similar combinations of tourism resources and environmental elements do not have the same structure as tourists, and some of the decisions affecting choice have been shown to be realized by influencing the establishment of satisfaction or loyalty. Mowen showed that tourists who develop a place attachment to a tourist destination will be more actively involved in tourism activities during tourism, and will make efforts to understand and learn about local customs and culture, thus creating a higher level of satisfaction with the tourist destination [31]. Alexandris found in an empirical study that ski tourists' place attachment has a significant positive effect on their loyalty [17]. Atila found that the degree of tourists' place attachment affects the evaluation of the destination and the establishment of loyalty, which is stronger than that of other destinations [19], using the example of Didim as a vacation destination, found that the degree of a tourist's place attachment affects the evaluation of the destination and the establishment of loyalty and that stronger place attachment positively affects loyalty and then tourism behavior [23]. Tsung Hung Lee, using the theory of social exchange, found that residents' perceptions of benefits from tourism impacts play an important mediating role in the positive influence of place attachment on the support for tourism development [3] Oliver showed that tourists receive a higher level of support for tourism development than those who receive a higher level of support for tourism development. showed that tourists obtaining higher satisfaction contributes to higher perceived value, which results in positive and positive post-trip behavior [21].

Of course, there are different research results on the relationship between place attachment and post-trip behavioral intention, George concluded that tourists' past travel experiences help them to form higher place attachment, and tourists' place attachment to the tourist place has a significant positive effect on tourists' next revisit [28]. Yu Yong in a study on the relationship between tourists' place attachment and post-trip behavior, pointed out that tourists' place attachment is mediated by value perception and experience quality, and has an impact on post-trip behavioral intentions, including the intention to revisit and recommendation intention [22]. Place attachment measurement adopted the consensus of Williams and Vaske's three-dimensional approach using the three dimensions of place identity, place dependence, and social bonding [23].

2.3 Tourism Behavioral Intention and Revisits

Measurement of tourism behavioral intention adopts the practice of scholars such as Cronin, Brady and Hult [34]; and Chen, and Tsai which is divided into two dimensions: repeat tourists' intention and recommendation intention. In tourism behavioral intention, revisit behavior is increasingly important for tourist destinations to gain market share, attracting a revisited tourist is easier and more effective than attracting a first-time tourist, and the cost will be reduced accordingly, Haywood found from a quantitative point of view, attracting a revisited tourist requires one-fifth of the marketing cost of a new tourist [34] Tourist attitudes towards revisiting are also intriguing and show a polarized tendency, with Baloglu and Erickson finding that some Mediterranean international tourists tend to visit a new destination the next time they travel, while others indicate that they will revisit the same place at some time in the future [27] Whereas the probability of revisiting behavior occurring is higher in destinations where the main function is leisure and vacation, Gitelson and

Crompton found through their study that the development of such as seaside, resorts and theme parks relies heavily on revisiting tourists [28].

In terms of the influencing variables of revisit intention, many factors have influence on revisit intention, which have been proved by research, including satisfaction, perceived value, and place attachment, in addition to the antecedent variables such as perceived image of the destination, tourism motivation, etc. The research results of Jang and Feng show that satisfaction is the direct antecedent of short-term revisit intention, and novelty-seeking motivation is the important antecedent of medium-term revisit intention. The important antecedent of medium-term revisit intention [29]. Baker and Crompton used structural equation modeling to show the positive influence between perceived value, satisfaction, and behavioral intention by empirical analysis [30]. Liu Li and Wu Hui selected Korean tourists as the research object and used Jiuhua Mountain as the research case site and verified that behavioral intention was influenced by tourism motivation through the method of empirical research [31].

2.4 *Lack of Empirical Tests of Role Identity in the Tourism Research*

Role identity originates from the theory of symbolic exchange, which holds that roles reflect individual identity and social status, as well as the relationship between individual behavior and psychology and group behavior and psychology. Scholars at home and abroad have made useful explorations of role identity measurement, such as Guo Liyan [12] who developed the Teacher Role Identity Scale for the teaching profession, and Wang Xi and Zhao Yufang [4] who developed the Role Identity Scale for College Student Volunteers. Drawing based on the more consensual Cheney identity scale [32], the study determined that the role identity scale contains four dimensions: self-identity, role perception, role emotion, and role behavior. Domestic tourism role identity and place attachment research is still in the exploratory stage, involving less research on the integration of tourism role identity, place attachment, and perceived value, especially the lack of empirical research to quantitatively analyze the mechanism of the relationship between the three and decision-making behaviors, using ski tourists as a case study.

2.5 *Literature Review*

When tourism development is examined more from the perspective of products, the advantage of customer value in the consumption process for the assessment of the aspects of paying costs and obtaining benefits becomes more obvious. When travelers choose a destination, it does not depend entirely on the combination of tourism resources and environmental elements, but the influence of psychological preferences is obvious, and antecedent variables such as travel motivation, satisfaction, loyalty, and of course place attachment may be important factors in the final choice of destination trade-offs. The influence of these factors on post-tour behavior has broken through the limitations of simply attracting tourists to visit a certain destination again, and more research has shifted to the attraction of tourists' feelings and emotions. From this perspective, it makes sense to introduce role identity into the tourism behavioral intention system, and it is even more necessary to verify its influence on the relationship between perceived value, place attachment, and behavioral intention, especially when it is more challenging to choose ski tourists in the hosting place of the 2022 Winter Olympics as the object of research. Therefore, the increasingly rich series of previous results not only laid a solid theoretical foundation for this study but also expanded the innovative space and technical path for the study of tourists' decision-making behavior system.

3 Methodology

3.1 Data Collection

Choose the 2022 Winter Olympics host Chongli ski destination as a case study, located in Zhangjiakou City, Hebei Province, belonging to the Beijing–Tianjin–Hebei synergistic development region, rich in ski resources, excellent natural environment, is a famous domestic ski destination. There are six ski resorts in the region, such as Yunding and Wanlong, with 107 snow trails, and the development of ski tourism industry agglomeration, the construction of international ski bases, and the creation of a collection of sports, vacation, and leisure in one ski tourism destination. In 2017, it received 2,741,000 ski tourists, and the tourism revenue was 1.96 billion yuan, a year-on-year increase of 6%. The group conducted a questionnaire survey of Chongli ski tourists between December 2018 and February 2019, 450 questionnaires were issued, and 432 questionnaires were recovered, of which 421 were valid questionnaires, and the validity rate of the questionnaire was 93.6%.

3.2 Measures

A structural equation conceptual model (Fig. 1) was constructed around analyzing the intrinsic mechanism of action between perceived value, role identity, place attachment, and decision-making behavior of ski tourists, with 13 latent variables and 46 observed variables.

Research hypotheses: based on the previous literature review, seven hypotheses are proposed. h1—Perceived value has a significant positive effect on decision-making

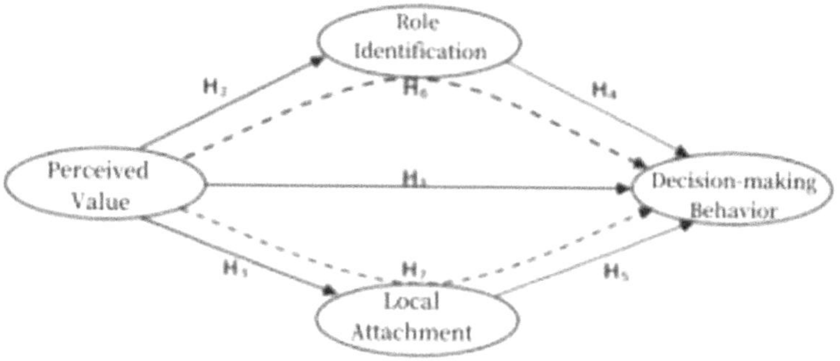

Fig. 1 Studies the conceptual model

behavior; h2—Perceived value has a significant positive effect on role identity; h3—Perceived value has a significant positive effect on place attachment; h4—Role identification has a significant positive effect on decision-making behavior; H5—Place attachment has a significant positive effect on decision-making behavior; H6—Perceived value is fully mediated through role identification and has a significant positive effect on decision-making behavior; H7—Perceived value is fully mediated through place attachment and has a significant positive influence. When analyzing the influence relationship between variables, the influence relationship of the embedded dimensions was analyzed at the same time.

3.3 Questionnaire Design

The questionnaire consists of five parts, the first part is the ski tourists' role identity scale, mainly in reference to Cheney [32] identity scale and Yang Shuo, Wang Dengfeng, etc. [33] on Rosenberg self-esteem scale validation, combined with the actual ski tourists to make appropriate revisions, with a total of 18 question items; the second part is the perceived value scale, mainly in reference to Pan [14] as well as domestic scholars Zhang Tao, Bao Jigang [15] and other scales, and combined with the actual part of the revision, a total of 14 items; the third part is the place attachment scale, mainly in reference to Williams and Vaske [23] and other scales based on the basis of the actual revision, a total of 9 items; the fourth part is the decision-making behavior scale, mainly borrowed from Cronin, Brady and Hult [24]; Chen [26] and other design scales, with a total of five question items; the fifth part is a survey of demographic characteristics of ski tourists, including gender, age, education, income, and skiing frequency, motivation, and other 11 question items. The first four parts of the question items were scored using a five-point Likert scale, i.e., a value of "5" means "strongly agree" with the listed viewpoints/expressions, "4" means "agree", "4" means "agree", "4" means "agree", "4" means "agree", "4" means "agree", and "4" means "agree". "agree", the value "3" means "not sure", value "4" means the value "3" means "Not sure", the value "4" means "Disagree" and the value "5" means "Strongly disagree". The content of the questions is in parts I–IV (Table 1).

4 Results

4.1 Descriptive Analysis

The demographic characteristics of Chongli ski tourists are as follows: 60.6% male, and 39.4% female, with male skiers predominating; the age span of 18–44 years old accounts for 78.6% of the total; skiing is similar to other sports, with young and middle-aged people as the main body; the education level covers a variety of levels, with undergraduate degrees accounting for 53.7% of the total, making the skiers more educated; the source of visitors is Beijing The majority of the source of 55.1%, followed by Hebei Province accounted for 26.8%, is the most important source of the market; high proportion of repeat trips, multiple trips accounted for 77.9%, the first trip only accounted for 22.1%; travel to relaxation, interest-based, accounted for 35.4%, 30.6%, respectively, to become the main motivation for skiing.

4.2 Structural Model Adaptability Test

Structural equation modeling fitness test, commonly used indicators are chi-square to degrees of freedom ratio (×2/df), regular fit index (NFI, CFI), goodness of fit index (GFI), adjusted goodness of fit index (AGFI), parsimony index (PNFI), value-added fitness index (IFI), and root mean square of approximation error (RMSEA, RMR). The questionnaire data were processed by applying AMOS 19.0 to obtain Table 2. Against the standard values, the test indices were passed, indicating that the construction of structural equation modeling fitness is acceptable [20].

4.3 Test Results and Analysis of the Structural Model

Combined with the AMOS26.0 path analysis test results (Fig. 2), the role paths of the three dimensions of perceived value and the four dimensions of role identity, the three dimensions of place attachment, and the two dimensions of decision-making behavior of ski tourists were further tested and analyzed (Table 3).

(1) The initial hypothesis H1 was not verified to be valid. That is, it means that there is no significant direct effect of perceived value on decision-making behavior for ski tourists.

Table 1 Potential variables and observed variables

variables	Items	Cronbach's α
Self-identification	RI1 can control lifestyle changes	0.808
	RI2 can experience happiness and joy from my family	
	RI3 Thinks people around me are accepting of me	
	RI4 knows exactly what type of person I am	
	RI5 can adjust myself in time	
Role recognition	RI6 group members are active for a common skiing goal	0.801
	RI7 The group's social image represents me well	
	RI8 Values are close to those of the group	
	RI9 I am proud to be a member of the group	
Role emotions	RI10 Can feel that the group cares about individuals	0.885
	RI11 With hard work, we can all meet the requirements set by the group	
	RI12 Have many common consumer ideas with other members	
	RI13 Would like to stay involved in the group's ski touring activities	
Role behavior	RI14 recommend the ski group to friends	0.857
	RI15 Will tell others about planned ski activities	
	RI16 considers the impact on the group when making skiing decisions	
	RI17 Often introduces themselves as a member of a ski group	
Social values	PV1Made a good impression on others during this skiing trip	0.866
	PV2Skiing has changed the way others perceive me	
	PV3Skiing has gained me social acceptance	
	PV4Skiing has earned me social prestige	
Emotional value	PV5Skiing makes me feel good	0.934
	PV6Skiing makes me feel relaxed	
	PV7Skiing makes me excited	
	PV8Skiing gives me a happy feeling	
Intellectual value	PV9I learned skiing skills from skiing	0.836
	PV10I learned the knowledge of ski equipment and outdoor self-rescue in skiing	
	PV11I have gained knowledge from skiing	
Functional value	PV12 I have enjoyed the excellent service in skiing	0.785
	PV13There are a lot of experiences to participate in here	
	PV14It is convenient to come here for skiing	
Local identity	PA1Chongli is important to me	0.876

(continued)

Table 1 (continued)

variables	Items	Cronbach's α
	PA2Love Chongli ski resort	
	PA3Strongly agree with Chongli Ski Resort	
Local dependence	PA4 For skiing, Chongli is the best choice	0.902
	PA5Satisfaction with Chongli ski resort is higher than other regions	
	PA6Skiing in Chongli is more important than in other ski resorts	
	PA7Wouldn't replace it with other ski resorts	
Social ties	PA8have many good memories of skiing in Chongli	0.810
	PA9Have a special feeling for Chongli and those who have visited here	
Willingness to revisit	BI1 I will ski in Chongli again in the next few years	0.901
	BI2 I would consider skiing here next time I have the chance	
Willingness to recommend	BI3 I would recommend skiing in Chongli to my friends and family	0.887
	BI4 I would like to introduce Chongli ski resort to my friends	
	BI5 Willing to introduce nearby tourist attractions to people around me	

Table 2 Structural model adaptability test

	×2/df	RMSEA	NFI	GFI	CFI	IFI	PNFI
Test value	2.317	0.056	0.958	0.945	0.976	0.958	0.748
Standard value	<3	<0.08	>0.9	>0.9	>0.9	>0.9	>0.5

(2) Initial hypotheses H2 and H3 were verified to be true. This indicates that the perceived value of ski tourists has a significant positive direct effect on role identity (path coefficient of 0.720) and place attachment (path coefficient of 0.717).

(3) Initial hypotheses H4 and H5 were verified to be true. This indicates that ski tourists' role identity has a significant positive direct effect on decision-making behavior (path coefficient of 0.135), and this direct effect is determined by the self-identity dimension, which has a significant positive direct effect on the willingness to revisit (path coefficient of 0.278) and the willingness to recommend (path coefficient of 0.161) in decision-making behavior. Place attachment also has a significant positive direct effect on decision-making behavior (path coefficient of 0.684), and this effect is due to the significant positive direct effect of place identity dimension and social bond dimension on willingness to revisit (path coefficients of 0.254, 0.138) and willingness to recommend (path coefficients of 0.289, 0.244).

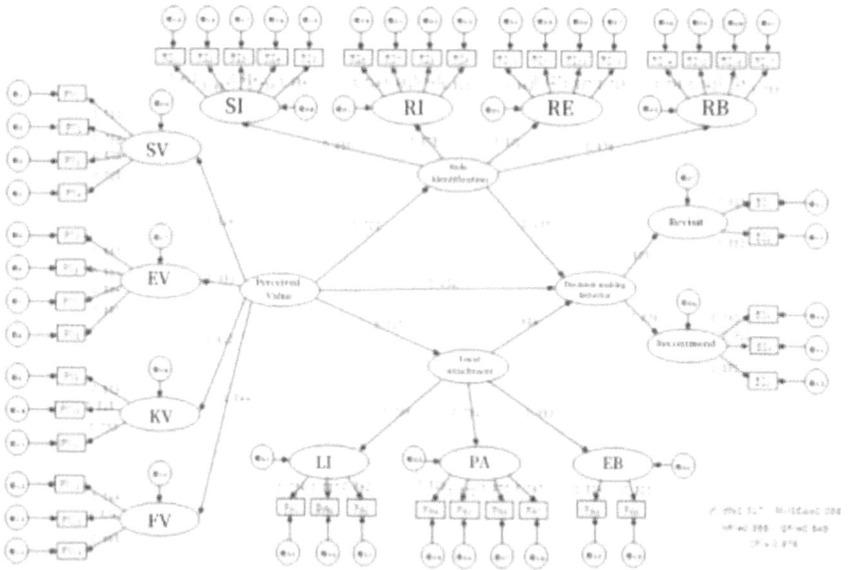

Fig. 2 Structural equation model of role identity, perceived value, place attachment and decision-making behavior

(4) The initial hypotheses H6 and H7 are verified by constructing the variable mediation effect test model. That is, perceived value has a significant positive direct effect on decision-making behavior, and when role identification and place attachment are added, the direct effect of perceived value on decision-making behavior becomes insignificant (path coefficient 0.137). Therefore, it is judged that there is a fully mediated effect of role identification and place attachment between perceived value and decision-making behavior for ski tourists.

5 Discussions

(1) The rationality of role identity as an antecedent variable of decision-making behavior. Role identity, as a psychological and social psychological variable, needs to be empirically tested when introduced into the category of tourism behavior. In this paper, it is found empirically that the role identity of ski tourists can directly affect their willingness to revisit or recommend, which indicates that role identity is an antecedent variable of decision-making behavior. The reason for this conclusion is that, compared with other tourists, the ski tourists' group has distinctive qualities, which are out of the high recognition of the role they carry or play in the group, and then produce more reasonable decision-making behavior.

Table 3 Model evaluation indicators and test results

Path			Standardized path coefficients	C.R	P
Perceived value	→	Role identification	0.720	6.442	***
Perceived value	→	Place Attachment	0.717	8.815	***
Perceived value	→	Decision-making behavior	0.132	0.873	0.147
Place attachment	→	Decision-making behavior	0.684	9.311	***
Role identification	→	Decision-making behavior	0.135	2.318	0.02
Perceived value	→	self-identification	0.768	10.826	***
Perceived value	→	Role recognition	0.713	10.005	***
Perceived value	→	Role emotion	0.803	11.384	***
Perceived value	→	Role Behavior	0.753	10.569	***
Perceived value	→	Place identity	0.75	10.936	***
Perceived value	→	Place dependence	0.665	9.669	***
Perceived value	→	Social bonds	0.811	11.847	***
Self-identification	→	Willingness to revisit	0.278	6.273	***
Self-identification	→	Willingness to recommend	0.161	3.897	***
Role recognition	→	Willingness to revisit	0.098	1.683	0.092
Role recognition	→	Willingness to recommend	0.061	1.12	0.263
Role emotion	→	Willingness to revisit	0.032	0.572	0.567
Role emotion	→	Willingness to recommend	0.013	0.245	0.806
Role behavior	→	Willingness to revisit	–0.05	0.855	0.393
Role behavior	→	Willingness to recommend	0.003	0.048	0.962
Place identity	→	Willingness to revisit	0.254	4.771	***
Place identity	→	Willingness to recommend	0.289	5.824	***
Place dependence	→	Willingness to revisit	0.052	0.931	0.352
Place dependence	→	Willingness to recommend	0.093	1.781	0.075

(continued)

Table 3 (continued)

Path			Standardized path coefficients	C.R	P
Social bonds	→	Willingness to revisit	0.138	2.226	0.026
Social bonds	→	Willingness to recommend	0.244	4.204	***

(2) Place attachment and role recognition can be important mediating variables for tourists' perceived value and decision-making behavior. In the paper, we adopt the structural equation method, based on the research of previous scholars, and try to introduce the role identity as an intermediary variable together with the place attachment, to construct a more systematic model of the influence of ski tourists' decision-making behavior. The empirical test shows that the structural model has strong rationality. The research results enrich the mediating system of decision-making behavior with place attachment as the core, and at the same time correct the conclusion that perceived value has a direct influence on decision-making behavior.

(3) For ski tourists, there are inter-dimensional differences in the significant direct influence of place attachment on decision-making behavior. The study shows that among the three dimensions of place attachment, place identity, and social ties have a significant direct impact on decision-making behavior, which in turn determines that place attachment is the antecedent variable of decision-making behavior; while the dimension of place dependence does not have a significant direct impact on decision-making behavior, which partially corrects the conclusions of Yu Yong, Tian Jinxia, et al. [12].

(4) Different types of ski destinations may have differences in the mechanism of action between their tourists' perceived value, role identity, place attachment, and decision-making behavior. Chongli is a hilly and mountainous ski destination with the logo of the Winter Olympics hosting place, and Beijing, Tianjin, and Hebei ski enthusiasts occupy the absolute majority, there is a very high revisit characteristic, so the research conclusion may apply to the source of concentrated and companion travel type of tourists. Northeastern mountain and forest type, urban snow-making type, and overseas ski destinations such as Europe, Japan, and South Korea should be selected as case studies for further empirical research on the mechanism of its role.

6 Conclusions and Further Discussions

This study is based on the fact that the consumer decision-making behavior can effectively predict and judge their consumption behavior. The research presented in this study is to determine the mechanism of the ski tourists' decision-making behavior. The study adopts the structural equation method and takes the Winter Olympics hosting place as the research object. It starts from the perceived value, role identity, place attachment, and decision-making behavior of ski tourists, to test the influence mechanism between the variables, and draws the following conclusions.

6.1 The Measurement Effect of Ski Tourists' Role Identity Scale is Empirically Tested.

Introducing role identity into the field of ski tourism, the biggest problem faced is the measurement of role identity, and the effect of this measurement needs to be tested accordingly in empirical evidence. At the first level, based on the systematic summarization of identity theories and the literature review, role identity was analyzed in terms of its cultural connotation, social structure, and behavioral characteristics in China, and the logical correlation and theoretical homology with self-identity and social identity were clarified. On the second level, we combined in-depth interviews to examine the individual and group role characteristics and dimensional composition of ski tourists, identified self-identity, role cognition, role emotion, and role behavior as the dimensions of role identity, and effectively integrated the content and structure of the test questions with the cultural connotation of Chinese tourists, educational characteristics, and the characteristics of the interviewees' words and behaviors.

During the exploratory factor analysis of Chongli as the research object, the reliability and validity of the Role Identity Scale were found to be good on the premise that the internal correlation between the survey items was good. In the process of constructing structural equation modeling and running it, all the fitness indexes meet the relevant standards and form a significant influence relationship between the variables. It should be said that the role identity scale has achieved the empirical test effect, which is not only the actual requirement for the operation of the structural equation model but also the rational basis for the introduction of role identity to construct the influence model of tourists' behavioral intention. However, it should also be pointed out that the exploratory study taken by the role identity measurement path is only verified in the empirical study of the Chongli area, and more empirical studies are still needed to test it for other ski tourism destinations.

6.2 Ski Travelers Show Obvious Group Characteristics.

Normally, gender, age, education, and income are selected as general control variables for tourism destinations, and due to the different types of tourism destinations and differentiation of tourism motivation, there are significant differences in the demographic characteristics of each variable.

In terms of the age difference of ski tourists, the significance coefficient value of age with all factors in the ANOVA is greater than 0.05, indicating that there is no significant difference in the factors. Although young and middle-aged ski tourists are the main ones, age has not become the main limiting factor of skiing, and compared with other types of tourist destinations, the source market of skiing destinations is broader, and it has unique charm and attraction for teenagers, young people, and middle-aged market. From the difference in the education level of ski tourists, the significant difference of each factor among ski tourist groups with different education levels found that ski tourists with college and bachelor's degrees pay more attention to factors such as role emotion and role behavior, which is also the most energetic group of ski tourists among Chinese nationals. From the perspective of behavioral intention, there is a significant difference between ski tourists with high school and lower education and those with college and bachelor's degrees, placing more importance on the intention to revisit.

In terms of differences in ski tourists' income levels, place attachment, and behavioral intention do not differ significantly depending on ski tourists' incomes, which suggests that the level of family income or even the affluence of ski tourists does not become a limiting factor for place attachment and does not play a decisive role in behavioral intention. From the gender differences of ski tourists, female ski tourists have a higher sense of self-identity, pay more attention to the functional value of scenic spots and the role of social bonding, and present a stronger willingness to recommend. Therefore, ski destinations can specifically target female ski tourists, improve, and enhance the function and value, to achieve the recommendation behavior to friends and relatives.

6.3 *Significant Differences in the Mechanism of Influencing Ski Tourists' Behavioral Intention*

Along the line of "cognitive-emotional-behavioral tendency", the study reconstructed the influence model of behavioral intention by introducing role identity and clarified the influence mechanism between perceived value, role identity, place attachment, revisit intention, and recommendation intention of ski tourists through empirical research. and willingness to revisit and recommend. Through structural equation analysis, hypothesis testing results confirmed that perceived value has a significant positive effect on role identity and place attachment, and that role identity and place attachment have a significant positive effect on behavioral intention.

The degree of influence of place attachment on behavioral intention is greater than that of role identity, indicating that place attachment has a more important role in influencing the behavioral intention of ski tourists. In the analysis of the relationship between the three dimensions of place attachment and the intention to revisit and recommend, it was found that the two dimensions of place identity and social bond had a significant positive effect on the intention to revisit and recommend, while the effect of the dimension of place dependence on the intention to revisit and recommend was not significant. As for the two dimensions that influence behavioral intention, the influence of place identity is higher than that of social bonds, and place identity becomes the most important factor for ski tourists.

Role identity has less influence on behavioral intention. Among the four dimensions of role identity, only self-identity has a significant positive influence on the intention to revisit and the intention to recommend, and the influence of the three dimensions of role cognition, role emotion, and role behavior on the intention to revisit and the intention to recommend is not significant. The statistical analysis of the questionnaire also shows that the mean value of the importance of self-identity of ski tourists is higher, which fully indicates that ski tourists pay more attention to this factor.

6.4 Mediating Effect of Role Identity and Place Attachment on Behavioral Intention

In the empirical process through the mediation effect analysis, role identity and local attachment on ski tourists' behavioral intention influence there is a mediation effect, that is, role identity and local attachment respectively in the perceived value and behavioral intention between the intermediary role, and local attachment mediation effect is more significant than role identity. From the composition of these two variables, whether it is the dimension setting or the specific content of the measurement items, role identity points more to the attitudes and tendencies maintained by the skiing group, and its influencing factors are mainly set around the skiing tourists; while place attachment focuses more on the emotion and judgment of the destination's functions, services, and the environment, and thus exerts a positive psychological influence on tourists, and it is precise because of these reasons that the degree of impact of place attachment on the behavioral intention of the skiing tourists is higher than that of role identity. It is for these reasons that the degree of influence of place attachment on ski tourists' behavioral intention is greater than that of role identity. Based on this, the current domestic ski tourism destinations can focus on building and increasing tourism projects that help stimulate the emotional factors of ski tourists' attachment, meet the individual and group psychological needs of tourists, and effectively guide the ski tourists' travel behavior and re-visit rate.

References

1. Bausch, T., Unseld, C.: Winter tourism in Germany is much more than skiing! Consumer motives and implications to Alpine destination marketing. J. Vacation Market. 24(3), 203–217 (2018). https://doi.org/10.1177/1356766717691806

2. Bausch, T., Gartner, W. C.: Winter tourism in the European Alps: is a new paradigm needed? J. Outdoor Recreat. Tourism 31, 1–9 (2020). https://doi.org/10.1016/j.jort.2020.100297
3. Correia, A., Pimpa˜o, A.: Decision-making processes of Portuguese tourist travelling to South America and Africa. Int. J. Culture Tourism Hospital. Res. 2(4), 330–373 (2008). https://doi.org/10.1108/17506180810908989
4. Denning, A.: From sublime landscapes to "White Gold": how skiing transformed the Alps after 1930. Environ. Hist. 19, 78e108 (2014). https://doi.org/10.1093/envhis/emt105.
5. Gardiner, S., Scott, N.: Destination innovation matrix: an empirical assessment. J. Destin. Manage. Market. 10, 122–131 (2018). https://doi.org/10.1016/j.jdmm.2018.07.002
6. Zeithaml, V.A.: Consumer perception of price quality and value: a means-end modell and synthesis of evidence. J. Market. 52(July):2–21 (1988)
7. Woodruff: Customer value: the next source for competitive advantage. J. Acad. Market. Sci. 25(2), 139–153 (1997)
8. Grewal, D., Monroe, K. B., Krishnan R.: The effects of price-comparison advertising on Buyers' perceptions of acquisition value, transaction value, and behavioral intentions. J. Market. 62(2):46–59 (1998)
9. Philip Kotler: Marketing management (10th ed.). Tsinghua University Press (2001)
10. Sheth, J.N., Newman, B.I., Gross, B.L.: Why we buy what we buy: a theory of consumption values. J. Business Res. (22):159–170 (1991)
11. Sweeney, C. Jillian, Soutar, N. Geoffery: Consumer perceived value: the development of a multiple item scale. J. Retailing 77(02):203–220 (2001)
12. Yonggui Wang, Hing PoLo, Renyong Chi, Yongsheng Y.: An integrated framework for customer value and customer-based perspective from China. Manag. Service Quality 14(213), 169–182 (2004)
13. Morrison, A. M.: Hospitality and tourism marketing. Albany NY, Delmar (1989)
14. Gallarza, M.G., Ruiz-Molina, M.E., Gil-Saura, I.: Stretching the value-satisfaction-loyalty chain by adding value dimensions and cognitive and affective satisfactions: a causal model for retailing. Manag. Decis. **54**(4), 981–1003 (2016)
15. Mowen A. J., Graefe, A. R., Virden, R. J.: A typology of place attachment and activity involvement: proceedings of the 1997 Northeastern Recreation Research Symposium. USDA Forest Service, Northeastern Forest Experiment Station
16. Alexandris, K., Kouthouris, C., Meligdis, A.: Increasing customers' loyalty in a skiing resort: the contribution of place attachment and service quality. Int. J. Contemporary Hospitality Manage. 18(5), 414–425 (2006)
17. Yuksel, A., Yuksel, F., Bilim, Y.: Destination attachlnent: effects customer satisfaetion and eognitive, affeetive and eonative loyalty. Tour. Manage. **31**, 27–284 (2010)
18. Tsung Hung, Lee.: Influence analysis of community resident support for Sustai able tourism development. Tourism Manage. 34, 37–46 (2013)
19. Oliver, R. L., Rust, R. T.: Service quality: new direction in theory and practice. Sage Publications (1994)
20. George, B. P., George, B. P.: Past visits and the intention to revisit a destination. J. Tourism Stud. 15(2), (2004)
21. Baloglu, S., Erickson, R. E.: Destination loyalty and switching behavior of travelers: a markov analysis. Tourism Analysis 2(2):119–127 (1998)
22. Gitelson, R.J., Crompton, J.L.: Insights into the repeat vacation phenomenon. Ann. Tourism Res. **11**(2), 199–217 (1984)
23. ang S C, Feng R M. Temporal destination revisit intention: the effects of novelty seeking and satisfaction. Tourism Manage. 28(2), 580–590 (2007)
24. Baker, D. A., Crompton, J. L.: Quality satisfaction and behavioral intentions. Ann. Tourism Res. 27(3), 785–804 (2000)
25. Cheney, G., Tompkins, P. K.: Coming to terms with organizational identification and commitment. Central States Speech J. 38(1), 10–15 (1987)
26. Huang, X., Deng, J., Ma, X., et al.: Spatiotemporal dynamics of snow cover based on multi-source remote sensing data in China. Cryosphere 10, 2453e2463 (2016). https://doi.org/10.5194/tc-10-2453-2016

27. Hjalager, A.-M., Kwiatkowski, G., Østervig Larsen, M.: Innovation gaps in Scandinavian rural tourism. Scand. J. Hospitality Tourism 18(1), 1–17 (2018). https://doi.org/10.1080/15022250.2017.1287002
28. Hallmann, K., Müller, S., Feiler, S.: Destination competitiveness of winter s port resorts in the alps: how sport tourists perceive destinations? Curr. Issues Tourism 17(4), 327–349 (2014). https://doi.org/10.1080/13683500.2012.720247
29. Li, Y., Zhao, M., Guo, P., et al.: Comprehensive evaluation of ski resort develop ment conditions in Northern China. Chin. Geogr. Sci. 26, 401e409 (2016). https://doi.org/10.1007/s11769-016-0808-z
30. Miragaia, D. A. M., Martins, M. A. B.: Mix between satisfaction and attributes destination choice: a segmentation criterion to understand the ski resorts consumers. Int. J. Tourism Res. 17(4), 313–324 (2015). https://doi.org/10.1002/jtr.2009
31. Pro¨bstl-Haider, U., Flaig, R.: The knockout deal: Pricing strategies in Alpine ski resorts. In: Pro¨bstl-Haider, U., Richins, H., Türk, S.: (Eds.) Winter tourism: Tren ds and challenges (pp. 116–145). CABI , Wallingford, Oxfordshire (2019)
32. Watanabe, M.E.: The United Nations sustainable development goals. Bioscience 70, 205e212 (2020).
33. Steiger, R., Abegg, B.: Ski areas' competitiveness in the light of climate change: comparative analysis in the eastern alps. In: D. K. Müller, M. Więckowski (Eds.) (2018)
34. Vanat, L.: International report on snow & mountain tourism: overview of the key industry figures for ski resorts. Retrieved from https://www.vanat.ch/RM-world-report-2019.pdf
35. Witting, M., Schmude, J.: Impacts of climate and demographic change on future skier demand and its economic consequences—evidence from a ski resort in the German Alps. J. Outdoor Recreat. Tourism 26, 50–60 (2019). https://doi.org/10.1016/j.jort.2019.03.002
36. Won, D., Bang, H., Shonk, D. J.: Relative importance of factors involved in choosing a regional ski destination: Influence of consumption situation and recreation specialization. J. Sport Tourism 13(4), 249–271 (2008). 10.1080/1 4775080802577185
37. Zach, F., Schnitzer, M., Falk, M.: Product diversification and isomorphism: the case of ski resorts and "me-too" innovation. Ann. Tourism Res. 90, Article 103267 (2021). https://doi.org/10.1016/j.annals.2021.103267

China Greater Bay Area Destination Image Perception by Foreigners Living in Guangdong

Jiangchuan Liang, Xialei Duan, and Huiyuan Dong

Abstract This study takes the Guangdong-Hong Kong-Macao Greater Bay Area as an example to explore the structure of intercultural interactions between foreign tourists and the tourism destination and see how it can impact the destination image perception. We performed a qualitative analysis of in-depth interviews with a group of foreigners living in Guangdong from 14 different countries. The study discovers that the foreign tourists interact with the tourism destinations in many critical encounters. It can be summarized in four different dimensions including material, society, institution, and symbol. They shaped the image of the interviewees' overall perception of this tourism destination, the China Greater Bay Area. As a result, an intercultural interaction structure model is constructed. We hope that the outcome and recommendations from this study can be helpful for the tourism destination image and strategy development in the post-pandemic era in this area.

Keywords Intercultural interaction · China Greater Bay Area · Foreign tourists · Destination image perception

1 Introduction

In 2019, 1.5 billion international tourist arrivals were recorded globally [1], and 133 million inbound tourists visited the Greater Bay Area of Guangdong, Hong Kong and Macau (hereinafter referred to as China Greater Bay Area) [2–4], both reaching record highs. However, in 2020, the COVID-19 pandemic broke out globally, and

J. Liang · H. Dong (✉)
Guangdong University of Finance and Economics, Guangdong, China
e-mail: 13691201005@163.com

J. Liang
e-mail: 22552662@qq.com

X. Duan
City University of Macau, Macau, China
e-mail: sallyduan@cityu.mo

Y. Luo et al. (eds.), *Tourist Experience and Perception in the Pandemic and Post-Pandemic Era*, https://doi.org/10.1007/978-981-96-1820-0_8

global international tourism fell sharply to a level comparable to that of the early 1990s [1], which made people rethink the sensitivity and vulnerability of international tourism. Although the recovery of international tourism has accelerated since 2022, as more countries and regions have lifted or further eased international travel restrictions, it will not be easy for international tourism to return to its pre-pandemic peak.

In the post-pandemic era, the changes in the global geopolitical environment caused by international instability factors, such as the worldwide economic recession, China-U.S. strategic competition, and the Russian offensive in Ukraine, have led to a series of problems for the global international tourism market, such as, weakened outbound spending power, low confidence to travel abroad, preference change of destination selection(more inclined to short-distance, well-infrastructural, safe and hygienic destinations), and increased thinking and attitude differences between destination residents and external tourists(hosts not welcoming foreign tourists for fear of virus infection, host–guest values, and ideological conflicts) [5–7]. All of these will profoundly affect the international tourism market directions and quantities, as well as tourists' behaviors in the coming period, and present new challenges to the intercultural management of international tourism.

The inbound tourism market, especially the foreign tourist market, is particularly important for China, not only because of the economic aspect of earning foreign currency and attracting investment but also in the light of the flexible "civil diplomacy" function of international tourism, which allows foreigners to know and understand the real China [6]. With a high level of economic development, openness and global accessibility, China Greater Bay Area is one of the major inbound tourism destinations in China as well as the first impression area for foreign tourists perceiving China [8]. Therefore, the continuous improvement of the perceived image of this area is very important to build China's national tourism image.

From the perspective of cultural interaction, international tourism is a kind of intercultural communication activity. There is no doubt that improving the tourist experience, particularly in terms of intercultural communication is an important approach of enhancing the international competitiveness of tourism destinations and promoting international tourism development. As transnational and transcontinental tourism has become more frequent, destination managers are paying more and more attention to international image building and communication, as well as the actual perceptions and real feedback of tourists of different nationalities on the destination image, thus, intercultural research on destination image perception has become a popular research area [9]. Due to the singular nature of international tourism, scholars have placed greater emphasis on intercultural communication strategies and the necessity for a nuanced understanding of how foreign tourists perceive destination images based on their Interactive experiences [10]. Tourists' perception of destination image based on the perspective of intercultural interaction is an important guideline for destination image optimization and marketing strategy improvement [11].

During the early stage of our study, a review was conducted related to cross-border tourism, intercultural exchange, and destination image, and look for research gaps. Then, semi-structured interviews were conducted on cross-border tourists using methods such as in-depth interviews, and the interview texts were analyzed. Based on the interview results, we found that cross-border tourists' perception of destination terrain is influenced by the interaction between tourists and multidimensional factors such as material, social, symbolic, and institutional factors. This study adopts a semi-structured in-depth interview method with foreigners who live in China Greater Bay Area, and attempts to explore how foreign tourists form their image perceptions of a destination through intercultural interactions. In view of the new changes in the international tourism market after the pandemic, this study proposes some strategies for building and promoting the China Greater Bay Area tourism image, which we hope to help to enhance the foreign tourists' travel experience and understanding of Chinese culture.

2 Literature Review

Due to the fact that tourists' perception of tourist destination imagery is the result of continuous interaction, in order to clarify the impact of cross-border tourists on the perception of tourist destination imagery, we have delved into how specific intercultural interactions affect tourists' perception of destination imagery.

2.1 Intercultural Interaction

Intercultural Interactions are usually defined as face-to-face contact between people from completely different cultural backgrounds [10], which is widely used in the fields of sociology and psychology. International tourism is essentially a micro-level experience of intercultural interaction [12]. Intercultural interaction is a dynamic process that changes with time and scene requirements [13]. There are three stages of intercultural interactive cognition, the first stage is intercultural perception, observation of other-ego cultural phenomena through the senses to obtain corresponding cultural impressions; the second stage is intercultural experience, through rich and diverse intercultural engagement and interaction, visitors are exposed to the deep roots behind cultural phenomena; the third stage, intercultural understanding, a process influenced by the interactive cognition of the self and the other, which takes the cultural nature of the self as a starting point and examines cultural similarities and differences to deepen understanding of the culture of the other [14, 15].

Intercultural interaction is widely used in multiple fields and can also be used to explain the perception and experience of cross-border tourists towards destination image. The intercultural interaction of foreign tourists is an important variable that directly affects tourist satisfaction and intercultural attitude [16]. Some scholars believe that the foundation of intercultural exchange in the tourism industry should be the recognition of cultural differences between international and local tourists. Dealing with intercultural communication in this way creates opportunities for both parties to recognize and value cultural differences, and experiencing cultural differences is one of the most fundamental driving forces for tourism and opening up reception communities to international tourists [17]. The intercultural competence of tourists represents their ability to interact effectively with people from different cultures, helping them gain a positive cultural exchange experience. Four factors have been identified for the intercultural competence of tourists: intercultural responsibility, understanding, appreciation, and action [18]. A study proposes criteria for measuring tourists' intercultural competence and establishes a framework to explain how cultural tourists interact with different cultures, thereby achieving unforgettable cultural experiences [18]. This study is the first to investigate the intercultural behavior patterns of long-term tourists to leisure vacation destinations with unfamiliar cultures. Four different levels of intercultural adaptation strategies were classified, with a greater focus on destination comfort and relationships building [19]. Exploring the issues of intercultural tourism and the role of English as the primary means of international communication [20].

Intercultural interactions in tourism are of multiple types and levels, and are not limited to the interaction between hosts and tourists, but also include the exchange of a full range of elements that influence image perception [20]. A study examined the structure and effects of intercultural interactions between international tourists and local residents, finds out that these interactions, through reflective, comparative, and comprehensive experiences, impact international tourists' attitude toward the local people and their culture [10]. For tourists, contact with local people can lead to a positive travel experience at the destination [24]. Therefore, social activities between visitors and hosts should be encouraged to foster mutual understanding and communication between them [25]. Scholars have conducted ethnographic discourse analysis on intercultural communication training courses for Nepalese tourism workers, their work abilities and skills can enhance their positive influence on tourists [39]. Hotel employees typically engage in frequent and intense intercultural service interactions with international tourists and customers, therefore these employees are expected to have a relatively high level of intercultural awareness and familiarity with intercultural characteristics and terminology [45].

To explore the mechanism of intercultural interaction, scholars have constructed corresponding research models from different perspectives. An integrated model of the intercultural interactions of international tourists from different countries, starting from shaping common themes, meanings, and patterns of international tourism experiences for participants [9]. A holistic psychological model was proposed, which includes the main contextual factors (cultural, situational, cognitive, and emotional contexts), to help better understand tourists' perception and evaluation of intercultural tourist guide service experiences, as well as the overall service quality of their travel experience [35]. Developmental Model of Intercultural Sensitivity (DMIS) was used to investigate the relationships between the background of volunteer tourism and changes in intercultural sensitivity [36]. Cultural adaptation model was proposed to examine the cultural adaptation strategies of cross-border tourism for immigrants and other travelers. Tourists face the task of maintaining their own culture while adapting to the host country's culture during international travel [41]. Exploring other places as tourists or short-term travelers is an opportunity for pleasure and self discovery, but it is also a confusing experience. A new theoretical framework—the dynamic model of cultural confusion—has been proposed to explain intercultural adaptation in tourism [42]. Beyond social exchange theory, combined with contact theory and comprehensive threat theory, a predictive model for tourist attitudes was tested. More frequent and satisfactory intercultural contact, lower threat perception, more positive stereotypes, and less intergroup anxiety will lead to positive attitudes towards the tourism industry among tourists [44].

2.2 Destination Image

Destination image is the combination of tourists' beliefs, opinions, and impressions toward the destination [26]. The process of cognitive processing of information is a process of encoding and decoding information from simple to complex, from low to high levels [24]. Image perception is a dynamic process of constant change, consisting of several materials that are stitched together, accumulated and processed by individual group consciousness, based on which they undergo several mental round-trips and filters [25]. The construction of a destination tourism image is a dynamic process of interaction between tourists and the destination [26]. Xie and Huang have summarized the five components of the elemental resource composition system of the image of a tourist place from the perspective of the tourist: the element of tourist resources, the element of tourist place facilities, the element of tourist place services, the element of industry management, and the element of community participation [27]. Kavaratzis proposed an interactive co-construction method for regional brands, namely, regional brand construction contains four basic elements: materiality, practices, institutions, and representations. These factors are the most direct factors that trigger people's associations with regions [28]. Jin et al. have crawled and analyzed the online comments of domestic and western tourists, and built a six-dimension framework of intercultural tourists' perceived destination image, which includes space, activity, interpersonal relationships, human-place relationships, time and affection [29]. In an research about Macau, the results indicate that tourists' destination image of Macau became enriched post-trip, after participants encountered the city and its people [30]. Martina G. Gallarza based on the existence of three dimensions of object, subject and attributes, previous studies are analyzed [31]. A content analysis was conducted with the aim of comparing the cognitive attributes of destination image and risk perception in the tourism field [32].

In summary, this study considers that the perception of destination tourism image is the result of the co-construction of foreign tourists and destinations, the interaction between tourists from different cultural backgrounds and the destination will have an important impact on the perception of the destination image. It is imperative for the China Greater Bay Area to form a strong synergy in building the tourism brand of the China Greater Bay Area from the perspective of overall destination image optimization. Therefore, this study attempts to put forward a generalized intercultural interaction theory of destination tourism image perception, study the contacts, interactions, and conflicts of foreigners in the process of tourism experience in China at the micro level, and explore the basic framework and key encounter points of intercultural interaction, which will be beneficial to enrich the theoretical system of intercultural interaction and have certain guiding significance for tourism image optimization and tourism marketing improvement.

3 Methodology

In order to understand the formation and influencing factors of cross-border tourists' perception of destination image, we designed an interview outline based on intercultural interaction theory and brand theory. Through in-depth interviews, we explored how tourists form and influence their perceptions of destination image through intercultural interaction.

3.1 Semi-Structured Interview

A semi-structured interview is adopted, and questions and answers are conducted according to the preset questions outline and open-ended questionnaire. The interview includes the cognitive elements of China Greater Bay Area tourism image and suggestions for China Greater Bay Area tourism development. Interviewers are composed of members of the research group and English translators. For native English speakers, interviewers communicate in English, while for non-native English speakers, interviewers communicate in Chinese. The interview duration of each interviewee is 30 min to 2 h. After the interviewee agrees, the interview is recorded, then converted into words, and proofread according to the results exported by translation software. In addition, after the interview, the members of the research group made participatory observations on the China Greater Bay Area tourist attractions and transportation hubs that foreign tourists often visit and listen to foreign tourists' evaluation of China Greater Bay Area tourism, to gain a more intuitive and in-depth understanding of tourists' perceptions of China Greater Bay Area tourism image.

3.2 Interviewees

Generally, foreign tourists' perception of a strange environment is superficial, so this study selected foreign interviewees who had lived in China Greater Bay Area for more than one year and had relatively rich travel experience, the selected interviewees included three types of groups: university students, English teachers, and foreign experts. The survey was conducted from April to June 2019. The interviewees were from 14 countries, including Korea, Japan, Russia, Indonesia, Vietnam, Thailand, India, Pakistan, Britain, Nigeria, Mauritius, Yemen, Ecuador, and Venezuela, with a total of 16 people, ranging in age from 20 to 60, of whom 4 were descendants of overseas Chinese, as shown in Table 1.

The interview in this study was divided into three stages. The first stage involved interviewing x individuals, the second stage supplemented the collected interview results, and finally conducted a second follow-up survey on the previous respondents. After three surveys, it was found that the results of the third interview could no longer extract new concepts, indicating that the interview questions in this study have been basically exhausted, indicating that the data has good comprehensiveness. From the perspective of regional distribution, it also basically represents the global regions, therefore, the data has good representativeness (Tables 2 and 3).

4 Data Analysis

This study conducts a qualitative study using rooting theory, imports the collated interview data into Nvivo12 plus analysis software for triple semantic coding, and then summarizes the influence of intercultural interaction on the perception of tourism destination image and its structure.

The first step is Open coding, which is a process of conceptualizing and categorizing the original interview data step by step. Its purpose is to identify phenomena, define concepts and discover categories. Firstly, the concepts are extracted from the interviewee's original discourse, and then the concepts are clustered according to the content similarity. Through two people and three rounds of screening and comparison, 200 short sentences are finally chosen, which provides a basis for the next step of code extraction.

The second step is Axial Coding. The purpose of this step is to analyze and explore the internal logical relationships between the categories acquired by open coding, and to integrate the scattered categories into a coherent whole again. The author further categorizes the 200 short sentences obtained from open coding based on their internal relationships and assigns concepts to them, thereby obtaining a higher-level category.

The third step is Theoretical Coding. Based on a comprehensive and systematic analysis of all relevant codes, the core categories interconnect through combinations or recombination, leading to the development of a systematic theoretical framework based on the correlations among categories. Thus, four dimensions of intercultural interaction of foreign tourists, namely, materiality, social, institutional, and symbolic are summarized, and the construction framework is shown in Table 4.

Table 1 Research models of intercultural interactions

Research model	Description	Authors	Year
Integrated model	International tourists shared themes, meanings and patterns that shaped participants' international tourism experiences	Yoo J, Sohn D	2003
Holistic mental model process	Contextual factors (cultural, situation, cognitive and affective contexts) to help better understand tourists' perception and evaluation of intercultural tourist–tour guide service encounters and the service quality of their overall travel experience	Zhao D, Y. Lin I	2014
Acculturation model	Tourists confront the task of maintaining aspects of their home culture while adapting to a host culture during their international travel	Rasmi S, Ng S I, Lee J A,	2014
Dynamic model of culture confusion	Exploring other places as a tourist or a short-term sojourner is a chance for pleasure and self-discovery, but also a confusing experience	Hottola P	2004
Predictive model of attitudes toward tourists	Combining the contact hypothesis and the integrated threat theory, more frequent and satisfying intercultural contacts, lower threat perceptions, more positive stereotypes, and less intergroup anxiety will predict tourists' positive attitudes towards tourism	Ward C, Berno T	2011
Developmental Model of Intercultural Sensitivity (DMIS)	Postulating that one's journey towards greater intercultural sensitivity consists of six consecutive stages along the ethnocentrism–ethnorelativism continuum, volunteer tourism is linked to positive and negative changes in intercultural sensitivity and therefore has the potential to simultaneously promote and inhibit intercultural understanding	Kirillova K, Lehto X, Cai L	2015

Table 2 Research object of intercultural interactions

Research subject	Description	Authors	Year
Host–guest	Local people can lead to a positive travel experience at the destination	Li Y Q, Liu C H	2020
	Social activities between visitors and hosts should be encouraged to foster mutual understanding and communication between them	Fan D X F, Zhang H Q, Jenkins C L	2017
	Interaction of tourists with local residents and tourism practitioners has a positive impact on the cognitive, emotional and symbiotic image of the destination image of the tourist	Stylidis D	2022
Tour guide-tourist	Tourists evaluating intercultural tourist–tour guide service encounters and the service quality of their overall travel experience	Zhao D, Y. Lin I	2014
Host community-tourist	Quality of interaction with the host community was the most significant predictor of change in intercultural sensitivity	Kirillova K, Lehto X, Cai L	2015
Volunteer-tourist	Volunteer tourism is linked to positive and negative changes in intercultural sensitivity and therefore has the potential to simultaneously promote and inhibit intercultural understanding	Kirillova K, Lehto X, Cai L	2015
Tour guides-tourist	Their work competence and skills include learning some tactics of self-presentation to create a certain kind of persona so that the guides can enhance positive affect in tourists	Sharma B K	2018
Hospitality employees-tourist	Hospitality employees were exposed to frequent and intense intercultural service interactions with international tourists and customers, who were expected to have a relatively high level of intercultural awareness, and familiarity with intercultural characteristics and terms	Koc E	2021

5 Findings

From the egocentric perspective, foreign tourists perceive the destination image through interaction with material, society, institution, and symbol dimensions. A series of specific key encounter points will be critical indicators for improving the tourism image.

Table 3 Basic information of interviewees

No	Country	Overseas Chinese	Major	Status	Gender	Age (years)	Time in China (year)
1	South Korea	No	Landscape architecture	Professor	Male	65	10
2	Britain	No	English Education	Foreign teacher	Male	37	5
3	Japan	No	Japanese education	Foreign teacher	Male	40	5
4	Nigeria	No	Marketing	Student	Male	22	1
5	Pakistan	No	Business English	Student	Male	28	1
6	Mauritius	Yes	Business Chinese	Student	Female	22	3
7	Indonesia	Yes	Business Chinese	Student	Male	22	3
8	South Korea	No	Business Chinese	Student	Male	22	2
9	South Korea	No	Business Chinese	Student	Male	23	2
10	Yemen	No	Business Chinese	Student	Male	22	2.5
11	Venezuela	Yes	Business Chinese	Student	Male	28	2
12	Russia	No	Tourism Management	Student	Female	25	4
13	Russia	No	Tourism Management	Student	Female	25	3
14	India	No	Tourism Management	Student	Male	36	5
15	Vietnam	No	Tourism Management	Student	Female	24	1
16	Thailand	Yes	Tourism Management	Student	Female	26	3

5.1 Interaction with Materials

The interaction between the tourists and materials refers to the direct contact and experience of foreign tourists at the level of material entities. Material interaction serves as the fundamental framework that molds visual perception. Key perceptual components encompass the natural environment, infrastructure, economic development and living consumption. According to interview findings, international travelers

Table 4 Theoretical coding, axial coding, and their material sources

Theoretical coding	Spindle coding	Source	Reference points	Reference point example
Interaction with Materials	Life consumption	10	37	1. In the China Greater Bay Area, only shopping is cost-effective, the accommodation and food are quite expensive 2. I think the tuition in Beijing is very expensive, but the tuition in Guangzhou is acceptable 3. In Hong Kong, where many tourists go to play and shopping, the censorship is very strict, so their products are good 4. In the Guangzhou Osmanthus Gang wholesale market, things are not expensive, and there is competition between booths, I know fakes are selling, but the quality is OK 5. Chinese wine is very popular, and people in our country like to drink it too
	Natural environment	6	16	1. The air quality in Guangdong is better than that in northern China 2. There is a coastline between Zhuhai and Zhongshan, which is also very beautiful 3. The first one I think is the weather, the weather is very hot, especially in Guangzhou, probably because the weather in my country (Ecuador) is not so hot 4. I've been to Dongguan and the pollution there is quite serious
	Economic development	6	14	1. China's economy is getting better and better, so now many people want to come here to do business 2. China is developing very fast. There are many big cities 3. There are a lot of tech startups in Shenzhen 4. Many Indians go to Foshan's furniture city to do business

(continued)

Table 4 (continued)

Theoretical coding	Spindle coding	Source	Reference points	Reference point example
	Infrastructure	5	7	1. Macau was a Portuguese colony with good infrastructure and European-style architecture 2. Hong Kong is very beautiful. The buildings are very high 3. In China it's cleaner around the cities because of the sweeper, but in Indonesia, it's less, so there's a lot of garbage and stuff
Interaction with Society	Language communication	7	33	Language has always been a barrier to travel and communication. We go to other countries, and speaking a little English can communicate, although we are not fluent, as you understand, I understand, it is OK. But how can you communicate when you don't even know the basics here?
	Trade & Exhibition	13	24	1. Guangdong is a commercial and trade center, and many foreigners come here mainly for business 2. I have a business plan after graduation. I think I can learn how to start a business here and establish some business relationships in Guangzhou
	Sports and recreation	10	13	1. If they choose to go to Macau or Hong Kong, Indians will prefer Macau because they like gambling very much and there is no gambling in India 2. I have a Chinese friend, he sometimes drives his car to take me to the surrounding cities to travel, I have traveled to many cities
	Transportation	5	11	Because China has convenient transportation facilities such as high-speed rail, the distance between cities in China Greater Bay Area has been shortened

(continued)

Table 4 (continued)

Theoretical coding	Spindle coding	Source	Reference points	Reference point example
	Daily social contact	5	10	1. After visiting so many countries around the world, I would think that the only country that dares to walk down the street after 10 PM is China. China is relatively safe, in Chinese cities, you can still go out at noon, but if I go out in Thailand, after 8 PM, it is dark, and I dare not to go out alone 2. People in Hong Kong are very polite, but here in mainland China, they are a bit impolite. If this characteristic does not change, how will we foreigners see China?
	Electronic payment	6	8	1. WeChat Pay and Alipay, are often used, very safe and wise. Every time I travelled to China before, I had to take a lot of cash because the bank didn't allow us to open an account for people with foreign nationality, so we had to bring cash. If you carry a lot of money, some people will stare at you and steal your money, and it is easy to be robbed
	Religious activities	4	6	1. Indians like primitive, Buddhist things, so the temples here are also attractive to Indians 2. Many foreigners, including Indonesians, Filipinos, French, and Germans, go to Ildek Road, Shamien Island, and Dongshan Pass to worship
	Part-time work	4	5	(Working as a part-time translator in the Canton Fair) The Canton Fair needs foreigners of different languages, there are many foreigners to come. When coming across foreigners from the same country as us, we can teach them how to go and help them buy tickets

(continued)

Table 4 (continued)

Theoretical coding	Spindle coding	Source	Reference points	Reference point example
Interaction with Institution	Visa facilitation	5	5	1. The tourist visa policies of Guangdong, Hong Kong and Macao are not convenient, and both mainland visas and Hong Kong and Macao visas are required 2. When we come to China, we have to make a visa, which is very complicated to submit to the government, and then the tourist visa only has a 30-day national period
	Inbound & outbound	4	6	What happens to a lot of people in Southeast 2021n countries is that the police pull you over and ask a lot of questions, even if you are a student, they ask everything, especially girls in Hong Kong
	International trade collaboration	5	11	1. Foreigners come to Guangzhou to buy raw materials, textiles, and resources and ship them back to their home countries 2. China and Thailand can exchange technology. For example, our country is backward in IT technology
	Tax free and Refund	2	2	Hong Kong and Macao have a tax exemption system, which is not available in other parts of China. This is a feature that can attract foreign tourists to shop here
	Foreign residency affairs	3	6	When it comes to talking directly to foreigners, for example, the responsible civil servants in the immigration department of the government, who speak English fluently, can explain common questions like visas, passports, etc
	Foreign media	1	2	There are few foreign media outlets in China Greater Bay Area, except Beijing-based China Daily, which has reported less in the area. If I lived in Hong Kong or Macau, I would know more about it

(continued)

Table 4 (continued)

Theoretical coding	Spindle coding	Source	Reference points	Reference point example
	Internet regulation	2	5	We can't use some apps, we have to open a VPN, and we have to pay to use that VPN. And then coming to China, and buying that SIM card, it's also a hassle, you have to use your passport, and it's expensive
Interaction with Symbols	Landmark scenery	12	42	1. The Hong Kong-Zhuhai-Macao Bridge is fantastic, I find it amazing, especially the bridge itself, which connects many cities and facilitates transportation 2. Macau is the ruins of St.paul, that's for sure, everyone will go there 3. On the map, there is a fence between Hong Kong and Macao, and the mainland. In reality, it is true
	Traditional customs	6	20	1. Hong Kong and Macao pay more attention to traditional culture 2. When the holiday comes, I will be exposed to some local traditional folk activities 3. Perhaps the most memorable part of Cantonese culture is martial arts 4. What attracts me most is the culture, especially traditional Chinese medicine 5. Sun Yat-sen is one of the greatest heroes of China
	Speciality	6	11	1. I think the best thing about Guangzhou is the food, dim sum, fish with pickled cabbage, and barbecue 2. Morning tea, Dude's dim sum, red rice sausage
	Local dialects	3	7	Cantonese is spoken as the first language in Guangdong, Hong Kong, and Macao

give precedence to factors like climate, air quality, drinkable water, oceanic conditions, urban architecture, public facilities, major industries, product quality, consumer prices, and souvenirs when assessing a destination. The natural environment notably influences individuals' initial impressions of a region where diverse interactions occur. Specifically, foreign tourists concentrate on the climate, air and water quality, sea conditions, and pollution within the China Greater Bay Area, shaping their judgments through personal perceptions and comparisons with their home countries. As for the material surrounding, the China Greater Bay Area, recognized as one of the most advanced regions in China, exhibits rapid and robust economic development making it highly appealing for business ventures. Therefore, it is used as a business tourism destination by a large number of tourists. In their daily consumption activities, foreign tourists emphasis on commodities, prices and quality. Importantly, their evaluations of material factors tend to be relatively objective and less influenced by subjective feelings.

5.2 Interaction with Society

The interaction between the tourists and inhabitants refers to host–guest contact in daily life, which is the most important part of intercultural interaction and has the most important influence on the formation of local tourism image. Specific cognitive elements include language communication, trade and exhibition, sports and recreation, transportation, daily contact, electronic payment, part-time work, and religious activities. According to the interview results, the contacts between foreign tourists and inhabitants go throughout the whole travel process, such as eating in restaurants, hailing cars, asking directions, shopping, business negotiations, etc. Hong Kong and Macao residents exhibit a high proficiency in the English language, allowing them to communicate fluently with foreign tourists. However, the prevalence of English in Guangdong lags behind that of typical tourist destinations. Currently, there are few host–guest interaction points that impress foreign tourists and are directly related to tourism activities. Foreign tourists have only mentioned gambling entertainment, road trips with friends, community building, root-seeking tours, and other activities. This indicates that there is still significant potential for enhancing opportunities for host–guest interactions in tourism activities. It is noteworthy that the widespread use of smartphone apps, such as WeChat, Alipay, Taobao, Didi, 12,306, and Baidu Translate, has also become an essential part of foreign tourists' perception of the local tourism image.

5.3 Interaction with Institutions

The interaction between the tourists and government regulation includes visa facilitation, inbound and outbound, international trade cooperation, tax free and refund, foreign resident affairs, foreign media and internet regulation, etc. When foreign tourists first arrive in China Greater Bay Area, national or regional border ports, such as airports, high-speed railways, and harbors, serve as their initial impressions of China. Foreign tourists assess China's national image and civilization based on the efficiency of government procedures and the sense of equal treatment to foreigners. According to the interviews, due to the current separation of mainland visas and Hong Kong and Macao visas, foreign tourists are limited in traveling across the three regions. Furthermore, the control over the distribution of foreign media and internet regulation in mainland China to some extent restricts tourists' access to information and sharing, and the lack of information can blur the original image of the destination in the minds of tourists.

5.4 Interaction with Symbols

The interaction between the tourists and the symbol refers to the process by that foreign tourists perceive and understand meaning through words, images, sounds, gestures, tastes and other substances. Symbols are not necessarily things that foreign tourists have experienced or fully understood. Everything that can positively interact with foreign tourists can be regarded as symbols, such as landmark scenery, traditional customs, special cuisine, local dialects, etc. From the interview results, landmark sceneries have become the classic images representing destinations in tourists' minds because they usually appear in posters or videos of various tourist destinations. Traditional customs are the carriers of local unique culture, including traditional festivals, skills, celebrities and so on. In terms of local cuisine, international tourists can further their connection with the "City of Gastronomy" identity by savoring Cantonese dishes, Cantonese-style tea snacks, and a wide array of global cuisines. Cantonese serves as the primary language in the China Greater Bay Area. Even though foreign tourists may not comprehend Cantonese, they can discern the contrast between Cantonese and Mandarin through their intonation, thereby enriching their perception of the distinctiveness of the China Greater Bay Area.

5.5 Overall Image Perception

Through the intercultural interaction of material, society, system and symbol, foreign tourists have formed an image perception of China Greater Bay Area tourism. According to the interview text, using Python3.7 and UCINET 6 software, the tourist image of China Greater Bay Area perceived by foreign tourists can be visualized. Select 51 words with word frequency not less than 6 in the interview text, construct a co-word matrix vocabulary, and draw a semantic network diagram of foreign tourists' overall comments (see Fig. 1). This map can only describe characteristics, but can't explain causality. The size of a node indicates centrality, and the larger the node, the more connections it has than other nodes. The thickness of the connection represents the correlation strength of two nodes, and the thicker the connection, the stronger the correlation between nodes.

In general, foreign tourists hold more positive perceptions of China Greater Bay Area's tourism image. Keywords such as convenience, difference, culture, development, economy, business, work, daily life, city, environment, traffic, and aesthetics have a high degree of centrality. This indicates that the overall image of China Greater Bay Area to foreign tourists is characterized by convenient city infrastructure, developed society and economy, aesthetically pleasing urban environment, unique culture,

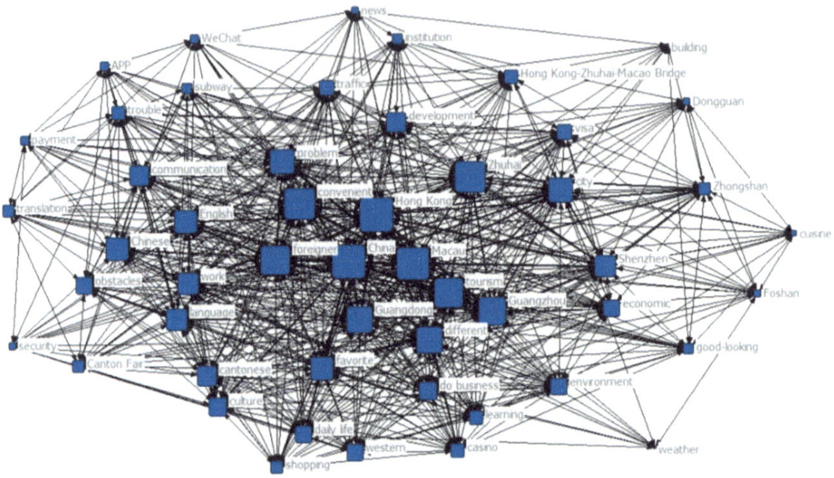

Fig. 1 Semantic network of foreign tourists' overall comment on China Greater Bay Area

and suitability for business, work, and life. It reflects the core traits of China Greater Bay Area, particularly contemporary, vitality, and cultural aspects. From the psychological perception mapping of the 11 cities in China Greater Bay Area, the spatial perception of foreign tourists in China Greater Bay Area is concentrated in the core triangle relationships of "Hong Kong-China-Macao", rather than the cities in Guangdong. Guangzhou, Zhuhai, Shenzhen, Zhongshan, Foshan, and other cities in Guangdong province exert relatively little influence on the perceptions of foreign tourists. Hong Kong, in particular, has emerged as a significant tourist destination due to its appeal as a shopping paradise and its proficiency in barrier-free English communication. Foreigners have rated Hong Kong positively, and it plays a central role in shaping the tourism image of China Greater Bay Area. Many foreign interviewees view their experiences in the China Greater Bay Area as reflective of life in China. Therefore, from their standpoint, the China Greater Bay Area encapsulates the quintessence of China.

From the perspective of cognitive content, foreign tourists pay more attention to factors such as city, language, economics, environment, building, institution, APP and cuisine, and emphasize focus on tourism, visa, payment, work, daily life, casino, do business, learning, shopping, transportation, translation and other activities. The scenic spot with the highest recognition is the Hong Kong-Zhuhai-Macao Bridge. From the perspective of semantic emotion, foreign tourists agree more with the traits of China Greater Bay Area such as convenience, culture, difference, development, security, western, aesthetics and so on, but they also hold a negative attitude towards language communication barriers and complicated visa application.

5.6 Integrated Model

Through the above research, this study proposes an intercultural interaction structure about how foreign tourists perceive China Greater Bay Area destination image (see Fig. 2) and confirms a series of encounter points of intercultural interactions on destination image perception, where positive interactions help to enhance destination image perception. Foreign tourists are self-centered and ultimately form an overall image perception of China Greater Bay Area destination through four dimensions of interaction experiences: material, society, institution and symbol. Intercultural interaction between tourists and destinations is a dynamic process that is infinitely derivable, and the various interactive encounter points provide a constant source of material and inspiration for the perception of the destination's image. For the inbound tourism market in the post-pandemic era, it will take some time to boost foreigners' confidence to travel abroad despite the lifting of entry requirements during the pandemic in various countries. Destination managers should make early preparations and mobilize stakeholders to strengthen the management of the whole services process before, during and after travel, and promote a good perception of the destination through positive interactions.

6 Conclusions and Implications

6.1 Conclusions

This study has identified the structure and key encounter points of foreign tourists' intercultural interactions in destination and summarized the impact of intercultural interactions on the destination image perception. The destination image perception of foreign tourists is a process of interacting with various key encounter points in four dimensions: material, society, institution, and symbol. This study expanded the application of intercultural interaction theory in the field of international tourism and provided a more comprehensive understanding of foreign tourists' perception of destination image.

In the post-pandemic era, the international tourism market faces many challenges, such as the need to enhance consumers' faith in the international tourism market, the multiple alternatives to international tourism, and the increasing risk of potential cultural friction between host countries and tourists. The results of this study will help the China Greater Bay Area destination marketing organizations (DMO) to develop effective marketing strategies and further improve the intercultural construction and communication of the tourism image of China Greater Bay Area and even that of China. Specifically, the following measures can be taken.

Firstly, China Greater Bay Area DMOs need to emphasize focus on intercultural interaction of foreign tourists, amplify foreign visitors' positive perception of China

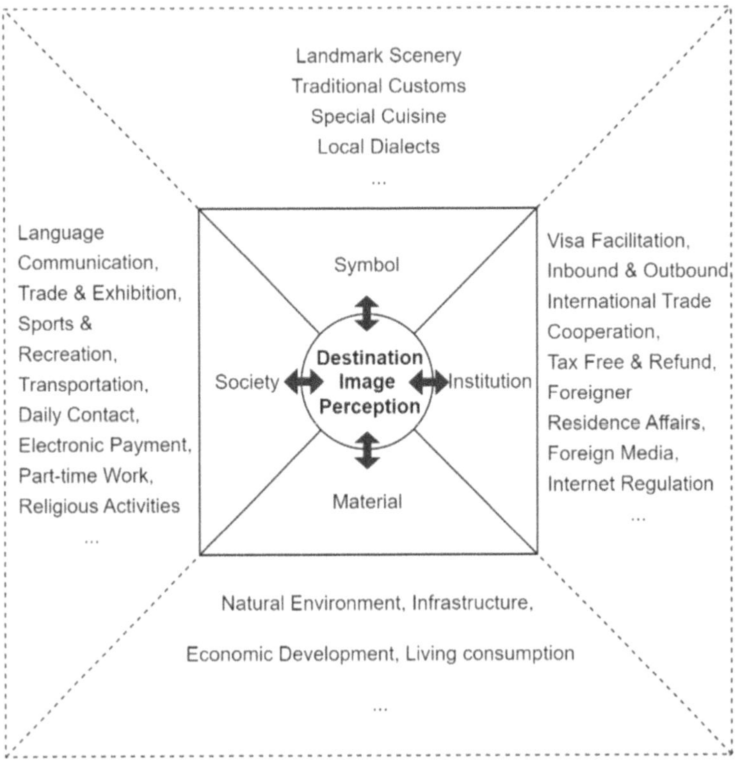

Fig. 2 Foreign tourists' intercultural interaction structure on China Greater Bay Area destination image perception

Greater Bay Area tourism, select visitor-perceivable interactive scenarios and activities based on the unique characteristics of China Greater Bay Area as a "contemporary China", "dynamic business metropolis" and "Canton culture", and promoting the brand image of China Greater Bay Area as a "world-class tourism destination". Secondly, it is important for tourism brand promotion and tourist market development to adhere to international practices, show respect for cultural differences, minimize conflicts and barriers to intercultural communication, enhance the process of intercultural interaction, and reduce negative perceptions arising from cultural differences. For instance, addressing the language communication issues, which were the most commonly complained about by interviewees, requires improvements not only in foreign language signage systems at key intercultural interaction sites but also in the language communication skills of tourism industry employees. Thirdly, DMOs need to innovate the digital marketing of destination image. Various daily life mobile apps have become an important channel for destination image perception, and the instant dissemination of user-generated content has accelerated the contact and communication between people of different cultural backgrounds. Foreign tourists are able to adapt the life well in China Greater Bay Area with the help of smartphone apps.

They are familiar with using daily life service apps such as WeChat, Alipay, Taobao, Didi Taxi, 12,306 railway booking, Baidu translation, etc. Therefore, DMOs can consider implanting tourism promotion content on apps commonly used by foreign tourists to build demand and attention for tourism in China Greater Bay Area in the minds of foreign tourists. Thirdly, DMOs should innovate the digital marketing of destination image. Various daily mobile applications have become an important channel for shaping destination image perception, and the swift dissemination of user-generated content has expedited interaction and communication among individuals from diverse cultural backgrounds. Foreign tourists can easily adapt to life in China Greater Bay Area with the aid of smartphone applications. They are accustomed to using everyday service applications such as Tik Tok, WeChat, Alipay, Taobao, Didi taxi, 12,306 Railway booking, Baidu translation, and more. Therefore, in the post-pandemic period, destination management organizations might explore integrating user-generated content apps for content marketing and precisely distributing tourism promotion materials combined with big data analysis to international tourists, in order to stimulate interest and demand for tourism in the Greater Bay Area.

6.2 *Limitations of the Study*

Some interviewees from non-English-speaking countries responded with less fluent Mandarin, leading to certain misunderstandings. Future research should try to conduct interviews with the participants in their native language. Conscious language expression is only a small part of subconscious thoughts. Future research should require continuous observation and long-term participation, and increase interview time and frequency. In particular, it is necessary to establish a familiar and mutual trust relationships between researchers and interviewees, to evoke more details of intercultural interaction in interviewees' deep memories. During the perception of destination image, foreign tourists may be influenced by various factors such as intrinsic values, demographic characteristics, behavioral patterns, cognitive abilities, and more. It is necessary to further investigate whether there are significant differences in the perceived destination image among tourists from diverse cultural backgrounds.

Note: This Paper is the Part of Result of thc Following Projects

1. Research on the Mechanism and Paths of Tourism Brand Co-Construction of the Guangdong-Hong Kong-Macao Greater Bay Area from the Perspective Of Host–Guest Interaction, supported by Humanities and Social Science Research Youth Fund Project Of Ministry Of Education Of China, 2020, NO. 20YJC790070.
2. Conservation and Development Strategies of Lingnan Water Town Traditional Village based on Landscape Gene Mapping, supported by Guangdong Province Social Science Planning Project, 2022, NO. GD22CYS18.
3. Innovation and Optimization of the Integration Mode of Culture Business and Tourism in the Protection and Inheritance of Lingnan Rural Cultural Heritage,

supported by Guangzhou Philosophy and Social Sciences Planning Project, 2022, NO. 2022GZGJ84.
4. Guangzhou Promotes Rural Revitalization and Urban–Rural Integration: Based on the Perspective of Innovative Development of Agriculture, Culture and Tourism Integration, supported by Guangzhou Philosophy and Social Sciences Planning Project, 2022, NO. 2022GZYB05.

References

1. UNWTO: International tourism growth continues to outpace the global economy, https://www.unwto.org/international-tourism-growth-continues-to-outpace-the-economy, 20 Jan 2020
2. Guangdong Bureau of Statistics. Guangdong Statistical Yearbook 2020, http://tjnj.gdstats.gov.cn:8080/tjnj/2020/directory.html
3. Hong Kong Tourism Board: Yearbook 2019, https://www.discoverhongkong.com/tc/about-hktb/annual-report/annual-report-20192020/index.html
4. Macao Statistics and Census Service: Macao Statistical Yearbook 2019, https://www.dsec.gov.mo/zh-CN/Home/Publication/YearbookOfStatistics
5. Bao, J.: The impact of the Covid-19 pandemic on tourism and reflections on future research [EB/OL]. Journal of Tourism 2020 Annual Conference on Tourism Research in China (2020), https://mp.weixin.qq.com/s/ncdhdi3pkNujt9Bhpe1aFA
6. Bao Jigang: Tourism should return to pre-pandemic patterns after May Day next year[EB/OL]. Guangdong Tourism Planning and Marketing Association (2022), https://mp.weixin.qq.com/s/Z4D6hFC_E3EgsEd11OyEqg
7. Xie, Chaowu, Lai, Fei-Fei, Huang, Rui.: Tourism resilience system construction and high-quality tourism development under the epidemic crisis. J. Tourism 37(09), 3–5 (2022)
8. Liang, J.C.: Research on brand personality perception of Chinese tourism cities. Guangzhou Jinan University Press (2016)
9. Jin, H., Huang, J., Sun, G.: An empirical study on the six-dimension model of tourist's perceived image: a case study of domestic and western tourists in Hangzhou West Lake. Res. Develop. Market. **36**(8), 807–814 (2020)
10. Yu, J., Lee, T.J.: Impact of tourists' intercultural interactions. J. Travel Res. **53**(2), 225–238 (2014)
11. Wang, W., Junyi, W., Amin, W.: The process of tourism destination image perception—an exploratory study based on rooting theory. Hum. Geogr. **33**(06), 152–160 (2018)
12. Reisinger, Y., Crotts, J.C.: Applying hofstede's national culture measures in tourism research: illuminating issues of divergence and convergence. J. Travel Res. **49**(2), 153–164 (2010)
13. MacKay, K.J., Fesenmaier, D.R.: An exploration of cross-cultural destination image assessment. J. Travel Res. **38**(4), 417–423 (2000)
14. Wang Zhiqiang.: Intercultural reception in cross-cultural hermeneutic perspective: cultural cognitive forms and cognitive assumptions. German Studies (1), 47–54 (2008)
15. Wang Zhiqiang.: The self and the other—cognitive interactivity of cultural similarities and differences in a cross-cultural germanic perspective. German Studies 2(21) (2006)
16. Echtner, C., Ritchie, J.R.B.: The measurement of destination image: an empirical assessment. J. Travel Res. **22**(4), 3–13 (1993)

17. Lee, K., Lee, N., Lee, T.J.: Korean tourists' experience of intercultural communication through English as a *Lingua Franca* (ELF) in overseas travel: a grounded theory approach. Tourism Manage. Perspect. **48**, 101145 (2023)
18. Yoo, J., Sohn, D.: The structure and meanings of intercultural interactions of international tourists. J. Travel Tour. Mark. **14**(1), 55–68 (2003)
19. Fan, D.X.F., Tsaur, S.H., Lin, J.H., et al.: Tourist intercultural competence: a multidimensional measurement and its impact on tourist active participation and memorable cultural experiences. J. Travel Res. **61**(2), 414–429 (2022)
20. Kirillova, K., Lehto, X., Cai, L.: Volunteer tourism and intercultural sensitivity: The role of interaction with host communities. J. Travel Tour. Mark. **32**(4), 382–400 (2015)
21. Fan, X.: The role of social contact between perceived cultural distance and travel attitude (2019)
22. Jingchuan Wang, Tianyuan Li.: Service Contact: a new perspective of constructing tourism public service system in destination. Tourism Tribune 27(03), 7–9 (2012)
23. Straub, J., Straub, J., Weidemann, A., et al.: Handbuch interkulturelle Kommunikation und Kompetenz. JB Metzler, Stuttgart, Germany (2007)
24. Jiuxia Sun.: Communication between host and guest and cultural Communication in Tourism. Tourism Tribune 27(12), 20–21 (2012)
25. Li, Y.Q., Liu, C.H.: Impact of cultural contact on satisfaction and attachment: mediating roles of creative experiences and cultural memories. J. Hosp. Market. Manag. **29**(2), 221–245 (2020)
26. Fan, D.X.F., Zhang, H.Q., Jenkins, C.L., et al.: Does tourist–host social contact reduce perceived cultural distance? J. Travel Res. **56**(8), 998–1010 (2017)
27. Crompton, J.L.: Motivations for pleasure vacation. Ann. Tour. Res. **6**(4), 408–424 (1979)
28. Matlin, M. W.: Cognitive psychology: theory, research and application. China Machine Press, Beijing (2016)
29. Xu Minghua, Jiang Kefan.: "Self-mapping" and image construction: the psychological formation mechanism of people's perception of the national image under the adjustment of the diplomatic pattern. J. Huazhong Univ. Sci. Technol. (Soc. Sci. Ed.) 33(06), 102–109 (2019)
30. Jiangchuan Liang, Shaohe Liu.: Research on factors and paths of tourism brand building in the Guangdong-Hong Kong-Macao Greater Bay Area. J. South China Univ. Technol. (Soc. Sci. Ed.), 21(05), 12–19 (2019)
31. Chaowu Xie, Yuanshui Huang.: On the participatory organization model of tourism destination image planning. J. Tourism (02):63–67 (2002)
32. Kavaratzis, M., Kalandides, A.: Rethinking the place brand: the interactive formation of place brands and the role of participatory place branding. Environ. Plan A **47**(6), 1368–1382 (2015)
33. Chen, Z.: A qualitative pilot study exploring tourists' pre-and post-trip perceptions on the destination image of Macau. J. Travel Tour. Mark. **36**(3), 330–344 (2019)
34. Gallarza, M. G., Saura, I. G., García, H. C.: Destination image: towards a conceptual framework. Ann. Tourism Res. 29(1), 56–78 (2002)
35. Khasawneh, M.S., Alfandi, A.M.: Determining behaviour intentions from the overall destination image and risk perception. Tourism Hospital. Manage. **25**(2), 355–375 (2019)
36. Anantamongkolkul, C., Butcher, K., Wang, Y.: Long-stay tourists: developing a theory of intercultural integration into the destination neighbourhood. Tour. Manage. **74**, 144–154 (2019)
37. Zhao, D. Y., Lin I.: Understanding tourists' perception and evaluation of inter-cultural service encounters: a holistic mental model process. Int. J. Culture Tourism Hospital. Res. 8(3), 290–309 (2014)
38. Gali, G.F., Shakhnina, I.Z., Zagladina, E.: International tourism as a substantial form of language interaction and intercultural communication. EntreLínguas **7**(7), 2 (2021)
39. Steiner, C.J., Reisinger, Y.: Enriching the tourist and host intercultural experience by reconceptualising communication. J. Tour. Cult. Chang. **2**(2), 118–137 (2004)
40. Sharma, B.K.: Training workers for intercultural communication in tourism. Language Intercult. Commun. **18**(4), 408–423 (2018)
41. Rasmi, S., Ng, S.I., Lee, J.A., et al.: Tourists' strategies: an acculturation approach. Tour. Manage. **40**, 311–320 (2014)

42. Hottola, P.: Culture confusion: intercultural adaptation in tourism[J]. Ann. Tour. Res. **31**(2), 447–466 (2004)
43. Stylidis, D.: Exploring resident–tourist interaction and its impact on tourists' destination image. J. Travel Res. **61**(1), 186–201 (2022)
44. Ward, C., Berno, T.: Beyond social exchange theory: attitudes toward tourists. Ann. Tour. Res. **38**(4), 1556–1569 (2011)
45. Koc, E.: Intercultural competence in tourism and hospitality: self-efficacy beliefs and the Dunning Kruger effect. Int. J. Intercult. Relat. **82**, 175–184 (2021)

Livelihood Sustainability of Tibetan Communities in a Heritage Tourism Destination—Jiuzhaigou, China

Xiurong Wei, Shuang Wu, and Li Cheng

Abstract Frequent disasters and crises in recent years not only threaten the safety of human life and property, but also directly or indirectly influence the conservation of world heritage and the sustainable livelihoods of tourism communities. Taking the Sustainable Livelihood Framework promoted by the UK's Department for International Development as a refernce, this paper explores the components and interactive mechanism of livelihood sustainability of tourism communities living at heritage sites. Taking post-earthquake Tibetan communities inside Jiuzhaigou World Natural Heritage Site as an example, this study finds that livelihood intentions are the endogenous power influencing residents' choice of livelihood strategy. As an important part of livelihood capital, cultural capital can enhance the resilience of community livelihood, alleviate the discomfort caused by shocks and externalize it into livelihood strategies. The research findings suggest that one of the main means of realizing the sustainable development of tourism communities in this particular site Jiuzhaigou is to promote its unique Tibetan cultural resources.

Keywords World heritage site · Livelihood sustainability · Tourism community · Cultural capital · Jiuzhaigou valley

1 Introduction

The tourism industry is dependent on natural environment and susceptible to changes in economy, politics and society, which makes it highly sensitive and vulnerable [1]. In recent years, major disastrous events frequented the earth, posing threat to

X. Wei · S. Wu (✉) · L. Cheng
Sichuan University, Chengdu, Sichuan, People's Republic of China
e-mail: 1072183796@qq.com

X. Wei
e-mail: gracewei@scu.edu.cn

L. Cheng
e-mail: chengli@scu.edu.cn

Y. Luo et al. (eds.), *Tourist Experience and Perception in the Pandemic and Post-Pandemic Era*, https://doi.org/10.1007/978-981-96-1820-0_9

the safety of world heritages. No matter it is disasters such as Hurricane Katrina, the Indian Ocean tsunami, and Wenchuan earthquake, or crises such as economic recession, political instability and the COVID-19 pandemic, they not only damage the safety of human life and property, but also directly or indirectly threaten the inheritance and protection of world heritage and the sustainable development of the livelihoods of communities at heritage sites [2]. Many World Natural Heritage Sites are located in those climate-sensitive, ecologically sensitive areas, and are prone to geological disasters. Once the world heritage sites are damaged, tourism development will be disrupted, posing threat to local economy and livelihoods of communities living in the heritage sites.

In face of sudden natural disasters, how can the tourism community located in a World Heritage site achieve long-term sustainable livelihoods? Sustainable Livelihood Framework (SLF) was originally promoted by the UK's Department for International Development (DFID) to address the issue of poverty and rural development. Since then it has been applied to the field of tourism and community studies [3–5]. Taking a people-centered approach, DFID's SLF has proven to be a useful tool for evaluating impacts of development initiatives and identifying potential strategies to improve livelihood sustainability [6]. Therefore, this paper introduces the theory of "sustainable livelihoods" into the study of post-earthquake livelihood of tourism communities in World Heritage sites, tries to use the grounded theory to build an analysis framework of sustainable livelihood of tourism communities in World Heritage sites after earthquakes, and takes the Tibetan tourism community in Jiuzhaigou Valley, a World natural heritage site with important influence at home and abroad, as an example.

This paper begins with a literature review, which traces the existing studies on heritage tourism and the community, post-disaster development of tourism community, and sustainable livelihood framework. Then in the methodology part, a post-positivist Grounded Theory approach is taken to examine sustainable livelihoods of post-disaster tourism communities, Jiuzhaigou World Natural Heritage Site is chosen as an example, and data collection and processing are elaborated. It is found that the sustainable livelihood framework of post-disaster tourism communities in Jiuzhaigou consists of "livelihood capital", "livelihood vulnerability", "institutions and systems", "livelihood intention", "livelihood strategy" and "livelihood outcomes". These findings are followed by discussions where the sustainable livelihood model of post-earthquake Jiuzhaigou tourism community is compared with the sustainable livelihood framework of DFID. Finally, countermeasures and suggestions are put forward in the conclusions.

2 Literature Review

2.1 Heritage Tourism and the Community

While the sustainable development of heritage sites is inseparable from scientific and effective management, local community plays an indispensable role in heritage management. World heritage sites attract the attention of tourists from all over the world for their natural landscape and cultural resources with outstanding universal value. They are home to community residents for generations. The community is not only part of a heritage, but also the main body which conserves and inherits the value of the heritage. To a certain extent, community residents "own" the heritage, and most communities participate in heritage tourism in some way [7]. It is important for heritage management to strike the balance between the sustainable development of heritage resources and community participation [8]. As main stakeholders, residents possess local knowledge and may provide insights into tourism conservation plans [9]. Community participation in heritage tourism can help residents gain confidence, self-esteem and pride through tourists' recognition of local natural and cultural values, stimulate their sense of ownership, and protect natural ecology and traditional culture [10].

The introduction of tourism induces changes in community livelihoods [11]. Instead of replacing original livelihoods, tourism can diversify existing economic activities, disperse risks of relying on one source of income, and enhance well-being, community identity and social cohesion [12]. According to Stone's study on the Chobe National Park and Chobe Enclave Community Trust, Botswana, tourism contributed to the community's financial capital by creating opportunities for self-employment through the formation of small business enterprises such as handicraft making and traditional dance groups, promoting infrastructure developments and service provision, and improving the community's political capital with the devolution of power over natural resource management [13]. In the case of Kibale National Park in western Uganda, tourism employment helped alleviate poverty, and diversify livelihood strategies by engaging in on-farm or off-farm activities [14]. For South Africa's Makuleke Contractual Park tourism project, local community benefited economically from creating small & medium tourism related enterprises and transforming community capital resources to economic multipliers [15].

Apart from those positive aspects, tourism development can also exhibits negative influence on community livelihood. First, the development of tourism may lead to forced migration of local communities, and the subsequent relocation would seriously disrupt the livelihood and survival of communities [16–18]. Second, relying solely on tourism is risky and may reduce the sustainability of community livelihood. As Lasso [19] showed in the case of Komodo village on Komodo Island, Indonesia, tourism development has led the local fishing community to give up their fishing to become dependent on selling souvenirs as one single source of income. They have abandoned their boats and relinquished their skills. Once tourism declines, it is impossible for them to return to fishing. Third, even if the community has gained development in income, employment and infrastructure, lack of community participation in tourism decision-making would limit development potential of livelihood [5]. Melubo and Lovelock [20] examined the impact of the world heritage site status on community livelihoods from the perspective of the Maasai indigenous community in the Ngorongoro Conservation Area in Tanzania. Compared with benefits brought about the WHS status and its related tourism, there was a low level of appreciation of the WHS among local residents. Most of them highlighted their discontent for inaccessibility to land resources, the destruction of existing livelihoods, unsatisfactory involvement in decision-making and benefits sharing, and negative attitudes towards relocation policies.

2.2 *Tourism Community and Disaster*

With the growing number of disasters, such as the outbreak of infectious diseases, earthquakes, terrorist attacks, nuclear explosions, financial crises and so on, tourist destinations around the world are also suffering from increasingly intense natural and man-made disasters [21, 22]. The development of tourism is closely related to the social and economic stability of natural environment. Any abnormal fluctuation in the social and geographical stability will shake the foundation of tourism development [8]. Due to the lack of resources and ability to rebound quickly, tourism cannot recover as quickly as other industries. Therefore the disaster brings a heavy blow to community and enterprises living in tourist destinations [2]. Disasters will not only destroy the natural environment and tourism resources, damage tourism service facilities, and affect the market order [23], but also affect the development of tourist destination communities [24], and restrict the sustainable development of tourism [22, 25]. As a result, scholars are paying more and more attention to the impact of disasters and crises on tourism and the development of post-disaster tourist destinations [26].

For disaster-hit tourism communities, residents are the most direct stakeholders of post-disaster development. The only way to achieve sustainable development of post-disaster communities is to turn passivity into initiative, and take actions to turn crisis into opportunity [27]. When external help is limited, community is the leading force of post-disaster tourism and plays an important role in protecting local culture and ecological environment [28, 29]. The more natural disasters the residents experience, the better their perception of tourism, the stronger the resilience of their communities [25]. Communities can build adaptive strategies that are compatible with the knowledge and capacity of local resources and acquire social capital, thus enhancing community resilience [30]. Research also found, benefit perception and community awareness can facilitate residents participate in post-disaster community response and enhance their ability to obtain social support in network and cope with external pressure [31]. Community attachment is a kind of psychological activity, including emotional cognition and behavioral intention, which is conducive to the positive response of the community to catastrophic events [32].

After the earthquake and tsunami in 2011, Japan began to develop a community-led post-disaster tourism development model, known as "blue tourism" [33]. Initiated by the community, Blue Tourism can effectively eliminate the tension in post-disaster reconstruction and provide useful insights into post-disaster community construction and cultural protection [34]. In addition, Kato [35] pointed out, understanding traditional culture and local knowledge contribute to the tourism development of post-disaster communities and the sustainable development of tourism. The transformation of disaster-affected communities into sustainable communities is an inevitable trend of post-disaster development.

2.3 *Sustainable Livelihood Framework and Tourism*

Originally aiming at addressing the issue of poverty and rural development, the concept of sustainable livelihood (SL) can be traced back to a report to the World Commission on Environment and Development (WCED, 1987), which pointed out that livelihood need to have strong resilience so as to achieve sustainable development. Then Chambers and Conway further elaborated on their understanding of SL, "A livelihood comprises the capabilities, assets and activities required for a means of living: a livelihood is sustainable which can cope with and recover from stress and shocks, maintain or enhance its capabilities and assets, and provide sustainable livelihood opportunities for the next generation" [36]. Although various scholars have given SL different definitions, in essence, SL is a mode of livelihood emphasizing sustainability of social development and ecological environment.

Based on the concept of SL, various agencies developed their sustainable livelihood approaches to address the issue of poverty elimination in various contexts, such as climate change, livelihood changes, and tourism development [5, 12, 37–40]. Among those approaches, Sustainable Livelihood Framework (SLF) developed by UK's Department for International Development [41] is extensively employed. This framework comprises five main factors that affect the sustainable development of livelihood: types of and access to capital, vulnerability context, transforming structures and processes, livelihood strategies and livelihood outcomes. Taking into consideration the elements of wellbeing, security and capability, vulnerability and resilience, and natural resource sustainability [26, 36], SLF offers a holistic and people-centered approach to evaluate the sustainability of community livelihoods [4, 39, 42]. It provides an analytical framework that helps build the understanding of livelihoods, and highlights the importance of understanding various livelihood components and factors, and reflects a dynamic process of multi-resource pooling and multi-sector cooperation [42].

Some scholars have taken SL approaches to explore the sustainable livelihoods of tourism communities. In their extensive studies on various communities in China, Su et al. found that tourism can effectively promote the transformation and innovation of backward livelihood modes in communities and improve the livelihood status of community residents in tourism destinations [43]. For example, in some coastal areas, tourism can replace the original livelihood such as fishing, and make up for the loss of income caused by the policy of fishing restriction. "Happy Fishing Family", a tourism form of experiencing the life of a fishing family, is developed to enrich the livelihood of coastal fishermen, and reduce the dependence of family livelihood on natural resources, thereby improving the local livelihood situation. It is also found that local knowledge and skills can be transformed into tourism livelihood capital, enhancing the sense of place attachment and pride of community residents while improving the travel experience [6, 43]. Besides, in the tourism context cultural assets could effectively transfer livelihood resources into livelihood strategies, thereby meditating the institutions and organizations [44]. Edwards et al. [3] integrated the cultural capital and the political capital within the Community Capitals Framework (CCF) into the SLF, and found that cultural assets such as languages, customs, ceremonies are crucial for the Waiapu indigenous community in New Zealand to maintain their unique identity. Srijuntrapun [45] employed SL approach to study urban world heritage, and found that lack of capital and community participation contribute significantly to an unsustainable livelihood, and in an urban context use self-reliance to measure livelihood sustainability seems inappropriate as people rely heavily on external forces. According to Aazami's study on Zarivar wetland in western Iran, the physical capital has the greatest effect on people's livelihood as it leads to the development of other livelihood capitals, and can directly improve financial, social and human capitals, participation of local people in conservation ensures fascinating nature experiences for tourists, and various conservation activities have helped in reducing livelihood vulnerabilities [46].

As can be noted from existing literature, natural disaster and crises frequented the earth in recent years, the sustainable development of heritage tourism and community livelihood are of vital importance. The concept of sustainable livelihood has promoted the construction of the theoretical system and analytical framework of sustainable livelihood in academia. Among those frameworks, DFID's SLF takes a people-centred approach and has been proved useful. It has been applied to various contexts, yet the sustainable livelihood of post-disaster tourism community living inside world natural heritage sites seemed to be little studied. Besides, for the study on heritage tourism and the community, very often most of the case studies were conducted in well-developed mature communities. For the study on the impact of disasters on tourism industry, most of the attention has been paid to post-disaster tourism destination planning, post-disaster tourism resources development, analysis of the impact factors and strategies of recovery and reconstruction. There is a lack of comprehensive and systematic research on livelihood sustainability after tourism destinations are affected by disasters. Based on literature review, this study would take a Grounded Theory approach to extract the core factors of the sustainable livelihood model for the disaster-affected tourism communities in Jiuzhaigou World Natural Heritage site and analyse the interrelation between different factors.

3 Methods

3.1 Grounded Theory

Developed by American sociologists Barney Glaser and Anselm Strauss in the 1960s, Grounded Theory (GT) gradually emerged into three main versions: Classical or Objectivist GT [47], Post-positivist GT proposed by Strauss and Corbin [48], and Constructivist GT suggested by Charmaz [49]. The current study employed Post-positivist GT paradigm due to the following three main reasons. First, GT can provide fresh insights into a phenomenon and establish theoretical propositions where little is known [47, 48, 50], while for this study sustainable livelihoods of post-disaster tourism communities inside world natural heritage sites is little studied. Second, GT is regarded as a more objective way of conducting qualitative research with its systematic coding of data [51]. It is appropriate for the examination of highly complicated phenomenon, and this study takes a holistic approach to evaluate the sustainability of community livelihoods which involves multiple sectors and numerous factors. Third, different from the other two versions of GT, the post-positivist coding paradigm comprises a preconceived analytical framework, and relies more on deductive than inductive reasoning [52]. This way of analysing the empirical data help researchers structure their data systematically and explore the connections between codes [53]. For this study on Jiuzhaigou Valley, empirical data would be analysed not only deductively based on the DFID's SLF, but also inductively to allow emergence of new concepts and factors, and intended to examine the interaction between different factors of SLF. Therefore, to adopt a post-positivist coding paradigm of GT would be most appropriate.

3.2 Study Area

Jiuzhaigou Valley is located in the Aba Tibetan and Qiang Autonomous Prefecture, northern part of Sichuan Province, China. With its vast karst land forms, spectacular waterfalls, and endangered plant and animal species such as the giant panda and Sichuan takin, Jiuzhaigou Valley was designated by UNESCO as a World Natural Heritage Site in 1992. There are nine Tibetan villages living inside the valley, namely Panya, Heijiao, Shuzheng, Heye, Yana, Zharu, Jianpan, Zezhawa and Guodu village, and Jiuzhaigou Valley is named after them. After the relocation of ethnic villages inside the valley in 2000, the original nine villages gradually merge into 3 communities, namely Shuzheng, Heye and Zharu community.

Influenced by Tibetan ethnic culture and religious belief, the communities in Jiuzhaigou are abundant in historical and cultural heritage. For Shuzheng Community, Shuzheng Village and Zechawa Village are located in the main tourist line. The villages are mostly decorated with colorful prayer flags and white towers. The external walls depict totems, and the interior decoration is the integration of Tibetan and modern styles. Heijiao Village is in the traditional Tibetan folk house style. Within Heye Community, Heye Village is located in the main tourist line, and the layout of the village is similar to that of Shuzheng Village. Yala Village and Panya Village are arranged according to the terrain, keeping the traditional Tibetan style of primitive simplicity. In Zharu Community, Guodu Village has the layout of traditional Tibetan village, while Zharu Village is a typical village pattern during the transition period between Tibet and Han. The archeological discovery of the Ashaonao Ruins on the east side of Jianpan Village is considered a large-scale settlement site of the Han Dynasty (Fig. 1).

With the progress of the time and the development of tourism, the whole Jiuzhaigou community has undergone great transformations. Jiuzhaigou is a place where the Anduo Tibetans, a branch of Tibetans in western Sichuan, have lived for generations. For thousands of years, the local residents have been leading a self-sufficient life with agriculture as the leading industry. Restricted by traffic, Jiuzhaigou is almost isolated from the outside world. This is probably the reason why the natural scenery of Jiuzhaigou has been well preserved. Until the 1980s, only a trickle of tourists entered.

The change of community livelihood in Jiuzhaigou Valley is synchronous with the development of community. Before the 1980s, community residents mainly lived on farming and animal husbandry. In 1978, Jiuzhaigou turned a nature reserve, which

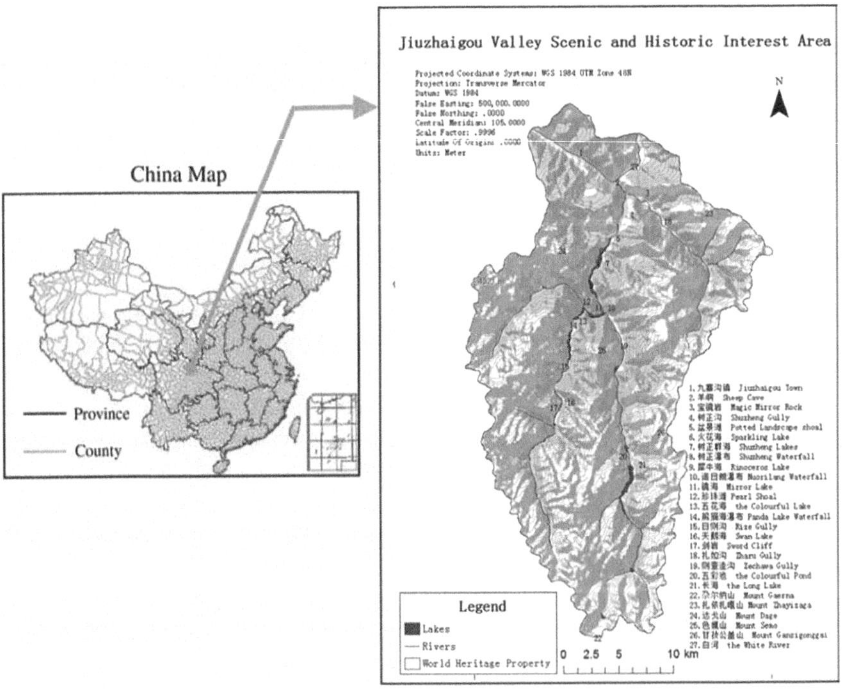

Fig. 1 Location of Jiuzhaigou Valley and three Tibetan Communities

attracted tourists to visit, so between 1980 and 1990, while agriculture still occupied the dominant position, tourism-induced sideline industry gradually developed. Residents began to engage in activities such as accommodation, catering, photographing, scenic spot management, and inspection. In 1991, tourism replaced agriculture as the leading industry, accounting for 80 percent of the workforce [53].

A major change for the livelihood of Tibetan communities in Jiuzhaigou is the implementation of a new policy "traveling inside the valley and living outside the valley" in 1998. To protect the world heritage, Jiuzhaigou Valley started to implement that policy, which strictly prohibited accommodation and reception activities inside the valley and relocated relevant businesses to places outside the valley. Residents of three local communities make a living by sharing part of the entrance fees with Jiuzhaigou Administration Bureau, selling souvenirs, providing horse riding service, loaning costumes with national characteristics to tourists for photo taking, as well as providing service outside the scenic spot [53]. Therefore, community livelihoods depend entirely on tourism development, with no other alternative industries.

Jiuzhaigou Valley is typical of those World Heritage Sites whose unique geographical environment contributes to its landscape with outstanding universal value, and yet they are prone to geological disasters and the ecological environment is fragile. On August 8th, 2017, A magnitude 7.0 earthquake jolted Jiuzhaigou Valley. The year before the earthquake, Aba Tibetan and Qiang Autonomous Prefecture attracted 37.6 million tourist with 31.8 billion yuan revenue. In 2017, tourism in Jiuzhaigou recorded negative growth for the first time in 5 years (−26%). The number of tourists decreased by about 8.5 million, and tourism revenue decreased by nearly 8.3 billion yuan. The earthquake not only caused great damage to Jiuzhaigou natural resources, but also a heavy blow to the tourism industry and the livelihood of the communities which depend on the World Natural heritage. To maintain the outstanding universal value of Jiuzhaigou World Heritage Site and the sustainable livelihoods for tourism communities living inside it became a problem.

3.3 Data Collection and Processing

The research group has been paying close attention to the development of Jiuzhaigou Valley, and immediately after the earthquake, the writer went to the disaster affected area to have a first-hand knowledge of the situation. Later the writer participated in meetings of Jiuzhaigou's post-disaster reconstruction, joined the discussion on the resilience of Jiuzhaigou, and traced the changes in community livelihood. For this grounded investigation into the sustainable livelihood of community inside the valley, a research group of ten went to the valley to conduct field work in June 2019. Qualitative research methods such as partcipatory or non-participatory observation and in-depth interview were employed so as to give interviewees enough time to think and express themselves. At the same time, interviewers could have time to observe the interviewees and try to understand their attitude, emotions and intentions.

In order to ensure the scientific representation and validity of the data, the interviewees in this study covered main stakeholders from almost all villages inside the Jiuzhaigou Valley such as indigenous residents, village cadres, people who move to Jiuzhaigou through marriage, shop owners and the staff of Jiuzhaigou Administration Bureau. During the interview, the writer communicated with the interviewee as a participant, tried to avoid any preconceived leading questions or hint, and recorded important information. All interviews were recorded and transcribed. Each lasted 30–60 min, while some key interviews lasted more than 1 h. After the interview, the author sorted out the interview recordings word by word and compared them with the notes made during the interview to ensure the accuracy of the interview texts. The sample number of this study was determined in accordance with the theoretical saturation principle. Finally 16 interview records and more than 80,000 words of interview texts were obtained. Each transcript was loaded to NVivo 11.0 software for coding (Table 1).

The coding process was divided into three stages: opening coding, axial coding and selective coding [48]. The coding process was divided into three stages: opening coding, axial coding and selective coding [48]. The initial coding stage comprised sentence-by-sentence analysis of the data which generated approximately 600 free nodes. Then after comparison and refinement, the free nodes were further selected and integrated into 81 initial concepts. During the axial coding, the conceptual nodes were further segregated into 24 subcategories based on their conceptual and theoretical

Table 1 Interviewee profiles

Interviewee	Village	Identity	Age	Gender
C01	Shuzheng Village	Resident	38	Female
C02	Shuzheng Village	Village cadre	41	Male
C03	Zezhawa Village	Resident	63	Female
C04	Zezhawa Village	resident	32	Male
C05	Zezhawa Village	Resident	60	Female
C06	Zharu Village	Resident	65	Male
C07	Zharu Village	Staff member of Jiuzhaigou Administrative Bureau	30	Female
C08	Rexi Village	Resident	31	Male
C09	Rexi Village	Resident	43	Male
C10	Zharu Village	Resident	27	Female
C11	Zharu Village	Monk	74	Male
C12	Zharu Village	Resident	65	Male
C13	Zharu Village	Resident	33	Male
C14	Panya Village	Village cadre	37	Male
C15	Valley entrance	Owner of shop	38	Female
C16	Jianpan Village	Staff member of Jiuzhaigou Administrative Bureau	40	Male

proximity. The coding process in grounded theory is circular. Along with extensive review of relevant literature, the data were constantly re-coded and emergent codes and categories were compared and redefined. Finally the link between the disparate subcategories was established and 6 categories emerged. Based on the categories derived from the initial and axial coding, the last stage extracted the core theme, "the sustainable livelihood of post-disaster tourism community". Around this core theme other main categories are connected, and a story line that could describe the whole process was drawn to form a theoretical framework.

4 Findings

After the three-stage coding process, six core elements of sustainable livelihood framework of post-disaster tourism communities in Jiuzhaigou were extracted, namely "livelihood capital", "livelihood vulnerability", "institutions and systems", "livelihood intention", "livelihood strategy" and "livelihood outcomes". The interview transcripts were reviewed together with the existing literature, and the connotation of the extracted core elements were identified.

4.1 Livelihood Vulnerability

For the livelihood of tourism community inside Jiuzhaigou World Natural Heritage site, livelihood vulnerability contains natural disasters and their effects, changes in the market trend, shocks brought about by events such as financial crises, ethnic conflicts, disease, and seasonality. Among them, natural disasters and their effects include earthquakes, debris flow, and landslides. These disasters will not only threaten the life and property safety of tourists and residents, but also damage the natural and ecological environment in the scenic area, lead to serious consequences such as damage to tourism industry, drop in family income, unemployment of community residents, and destruction of environmental facilities.

"C09: Now Panda Lake is the earthquake area. It is basically unable to go. It have been destroyed. This scenic spot may not be able to recover. The Nuorilang Lake and the Sparkling Lake have also been affected by the earthquake, but not as serious as the Panda Lake."

The seasonality of the tourism industry also restricts the sustainable development of livelihood. Holidays and the best viewing time of scenic spots bring obvious low and peak seasons of tourism, resulting in fluctuations in tourism income. Jiuzhaigou Scenic Area has obvious off-peak season before the earthquake, and the recovery of tourism after the earthquake should be adjusted reasonably by evaluating the market trend.

4.2 *Livelihood Capital*

Within the sustainable livelihood framework of tourism communities in post-earthquake Jiuzhaigou, livelihood capital consists of economic capital, social capital, human capital, natural capital, physical capital and cultural capital. Economic capital is necessary capital reserve, social capital is auxiliary resource, physical capital is the foundation, cultural capital is the guiding factor, natural capital is the guarantee of sustainable development, and human capital is the key to coordinate the utilization of the other five capital (Fig. 2).

4.2.1 Economic Capital

Before the earthquake, residents inside the valley had some family savings and improved their living standards by developing tourism. After the earthquake, outside enterprises and enthusiastic people actively donated money and materials to Jiuzhaigou to help residents through the darkest period. The state had also allocated a large amount of fiscal funds for post-earthquake reconstruction projects. For buildings damaged by the earthquake, assessment was carried out, some dangerous house were reinforced, and some were demolished and reconstructed. Apart from

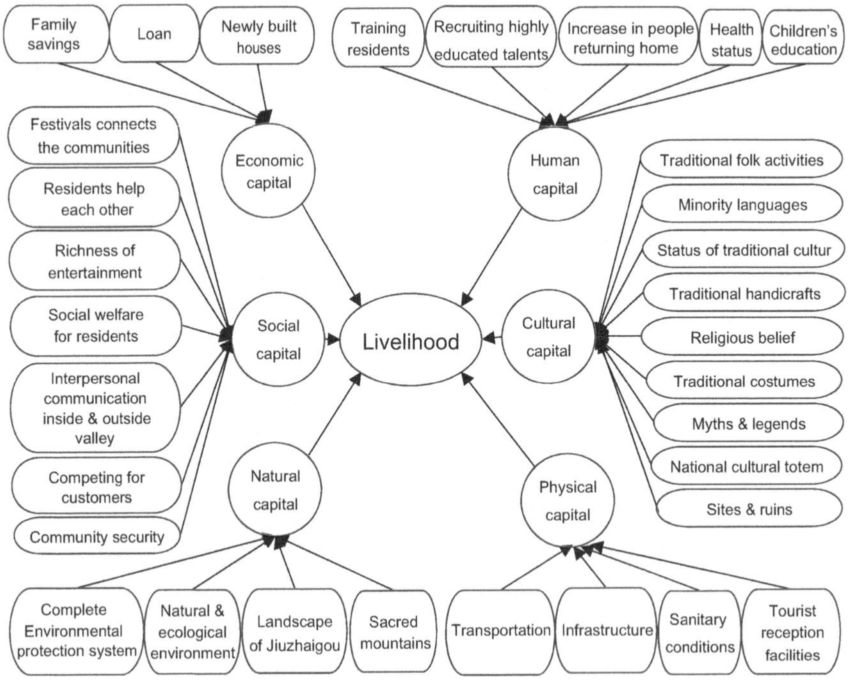

Fig. 2 Components of livelihood capital

that, the local village committee also took the initiative to support poor families affected by disasters and help them get loans.

4.2.2 Social Capital

In terms of social capital, the occurrence of disasters increases social capital. The geographical location of each village in Jiuzhaigou is different, resulting in the unequal distribution of benefit among community residents from tourism development. Phenomena, such as envy, competing for tourists, and disobeying management occur frequently. After the earthquake, the tourism industry was interrupted, and there was no interest involved between communities. The neighborhood relationship between communities was no longer as tense as before. Due to the disaster, residents gathered together again to comfort and help each other. At the same time, the disaster also aroused the national cultural values deeply embedded in the hearts of community residents, the mood of national unity was high, and the social capital was strengthened.

"C04: Before, for example, I had eight shops here, and he was not comfortable when I made more money than him... When individual visitors came, there would be lots of competition. For example (when tourists) went to his shop, (others may) shut down electricity, did what all him could to make you unable to sell... After the earthquake the situation became better. We have no business to do anyway. Before the earthquake, you had to fight for business. Now everything is transparent, (we) can gather together and have a chat. And we Tibetans are still very united. After the earthquake, when one family were in need, all other families would go to help. As for one family, one senior lived alone, his children were not around, other residents gathered to pay him a visit."

4.2.3 Physical Capital

Although the disaster caused damage to the ecological environment, infrastructure and industrial development, the post-disaster reconstruction undoubtedly became an opportunity for the redevelopment of Jiuzhaigou. The post-disaster reconstruction project inside the valley involves all aspects of the community, including transportation, water and electricity pipelines, cable network and so on, which is conducive to reshaping the local infrastructure and consolidating the foundation of physical capital. As for external transportation, Jiuzhaigou Airport, trains, expressways and other transportation modes are under accelerated planning and construction, striving to realize the rapid connection with Jiuzhaigou scenic area, and improve the accessibility of post-earthquake tourism. While for internal transportation, in addition to accelerating the restoration of damaged roads in the scenic area, tourist walkway is under intensive construction. The walkway connects the main scenic spots and villages in the scenic area, which not only reasonably diverts tourists, but also meets the needs of economic development for communities in remote area.

4.2.4 Human Capital

With the development of tourism in full swing, many youth in the valley who originally went out to work or study choose to return home and participate in tourism-related activities. Most of the younger generation have received college or higher education, and many have even obtained graduate degrees, further improving the education level of residents in the valley. After the earthquake, although the residents were temporarily unemployed, communities inside the valley organized the talent exchange, skill training and other activities to reserve high-quality service personnel for the development of post-earthquake tourism.

"C14: Our young people also choose to go out, mainly to study. During the post-disaster reconstruction, we still have counterpart support projects with many provinces and cities. For example, we have a government assistance program with Jiashan City, Zhejiang Province, which provides employment and training, covering round-trip air tickets, accommodation and so on."

4.2.5 Natural Capital

Although the earthquake damaged Jiuzhaigou scenic spots to a certain extent, as a world natural heritage site, Jiuzhaigou has a strict and complete ecological environment protection policy, which provides natural capital guarantee for the sustainable development of tourism communities after the earthquake. In addition, the fundamental principle of post-disaster reconstruction inside the valley is to focus on ecological protection and ecological restoration, and make every effort to maintain the integrity of the world natural heritage and highlight its universal value (OUV).

4.2.6 Cultural Capital

With its long history of Tibetan settlement, Jiuzhaigou World Natural Heritage site not only has unique beautiful landscape, but also has rich cultural resources, including ruins, folk customs, religion, traditional crafts, songs and dances and so on. Different communities have their own cultural resources. The Shuzheng community is dominated by its relics culture, with the Ashaonao Ruins and traditional Tibetan villages. Heye community stands out for its folk culture based on Tibetan weaving and ethnic festivals. Zharu community is famous for its religious culture centered around Zharu Temple, Zayizaga Sacred Mountain, and Baojing Rock. Therefore, for tourism development after the earthquake, the three communities can supplement the tourism resources with their unique ethnic and religious culture, which can enrich the tourism industry in the valley after the earthquake.

"C06: We all went to Benri Sacred Mountain in the recurrent year of our birth in the twelve-year cycle. It's the only way of walking around the mountain following the tradition of Ben cult. There are many relics in Benri Sacred Mountain, as there are on Zayizayunai Sacred Mountain. In the past, there were many caves for Taoist monks who practiced. This associates the Tibetan settlements with sacred mountains, temples and residents."

Cultural landscape, together with natural landscape, constitute the comprehensive scenery of Jiuzhaigou Valley, which is an important element to form the "place" characteristics of Jiuzhaigou, the essence and core characteristics of the tourism image of Jiuzhaigou, and the foundation for the tourism revival of the community after the earthquake. In the interview, many residents also expressed their strong desire to incorporate traditional culture into Jiuzhaigou tourism industry.

"C13: Jiuzhaigou culture is also very important to Jiuzhaigou tourism. Now no matter what is said, culture is always mentioned. In addition to see the scenery of Jiuzhaigou, visitors came to feel our unique local customs and practices...... In our village, not only the scenery can promote tourism in Jiuzhaigou, but also the Tibetan culture."

4.3 Institutions and Systems

Before the earthquake when Jiuzhaigou began to develop tourism, the market was not standardized and there was no third-party constraints, which lead to collective irrationality under the temptation of economic interests. Under that circumstance, the government began to standardize the management of Jiuzhaigou tourism market. The Jiuzhaigou Valley was jointly managed by the Aba Prefecture and the Jiuzhaigou County government, with Aba Prefecture playing the leading role, and Jiuzhaigou Administrative Bureau being in charge on behalf of them. The county government took charge after the earthquake. On the one hand, multiple management and unclear delineation of regulatory agencies' power and responsibilities lead to difficulties in solving the problems of residents and affect the livelihood and development environment. On the other hand, Jiuzhaigou Administrative Bureau provides jobs for residents and their children, mainly engaging in management, sanitation, forest fire prevention, forest administration work and so on. Excellent candidates can be promoted to management.

Jiuzhaigou community has formulated and implemented a number of policies that affect the sustainable livelihood of the communities in the valley. For example, the heritage site protection system and the national policy of returning farmland to forest change the identity of local residents from rural to urban, but the change does not bring corresponding benefits in social welfare. Instead, it restricts the residents from choosing diversified livelihood strategies. The policy of "travel inside the valley, live outside the valley" leads to the relocation of business activities, and the development of livelihood in the valley is blocked. The tourism income sharing system, including sharing revenue from scenic spot tickets and sharing the income from Nuorilang Tourist Service Center, makes residents lose the motivation to seek livelihood elsewhere, intensifies the dependence of residents on natural capital, and increases the vulnerability of livelihood. Accommodation reception inside the valley is strictly prohibited, which intensified the conflict between residents and the

Jiuzhaigou Administration Bureau. In addition, the county government also made construction drawings for new building without consulting the residents, and forced residents to build houses based on the drawings, which also caused opposition from most residents.

4.4 Livelihood Intentions

Through the analysis of the interview texts, this study introduced the concept of "intention" into the sustainable livelihood framework of post-earthquake tourism communities, and proposed that "livelihood intention" is the endogenous power influencing residents' livelihood strategy selection. Livelihood intention is composed of livelihood expectation, livelihood will, and ability and constraints. Livelihood intentions are brought by community residents' assessment of previous livelihood strategies, judgment on current external environment of livelihood, reflection on the advantages and disadvantages of its own livelihood ability, and expectation for future livelihood. This behavioral intention will affect the investment of capital livelihoods, choice of livelihood strategies, and finally livelihood output (Fig. 3).

On the one hand, community residents inside the valley are full of expectations and confidence for the future re-opening of Jiuzhaigou, and offer suggestions for the future development. They believe that incorporating Tibetan culture, Kangba culture and religious culture into Jiuzhaigou tourism and creating the product line based on cultural travel experience of Tibetan villages, Zayizhaga Sacred Mountain,

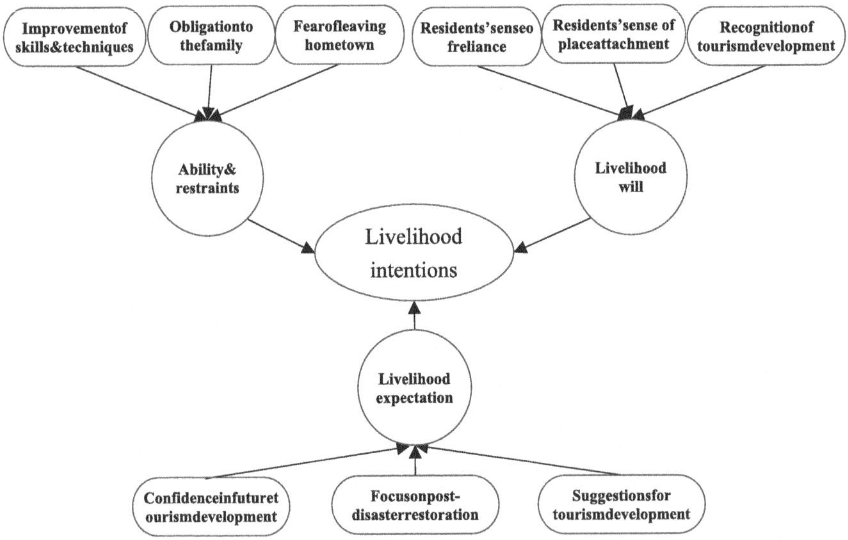

Fig. 3 Components of livelihood intentions

and Zharu Temple could add greatly to Jiuzhaigou tourism development after the earthquake. On the other hand, residents have expressed concerns about whether Jiuzhaigou can be restored as before, paying close attention to the post-earthquake reconstruction process. At the same time, by comparing and evaluating the livelihood results, residents will decide the proportion of investment in a new round of livelihood capital and the livelihood strategy to be adopted. In the interview, community residents generally reflect that the single livelihood strategy is not sustainable, and believe that in order to restore and even further develop the livelihood of Jiuzhaigou community after the earthquake, it is necessary to develop some non-tourism related livelihood activities in addition to tourism, and promote sustainable development of livelihood with the diversity of livelihood.

Before the earthquake, tourism in Jiuzhaigou has been developing well, and there are the senior and the young in the family to be taken care of, so the responsibility and obligation to the family is one of the factors for young people to return home to engage in tourism. In addition, the local authority's conscious efforts to train residents in skills and provide them with employment opportunities have further promoted the sustainable development of tourism-related livelihoods. Moreover, Jiuzhaigou has been living on tourism for more than 30 years. Community residents have a strong sense of reliance on the development of tourism. They are reluctant to move out of Jiuzhaigou because of their attachment to the hometown. Therefore, tourism should still be the starting point to solve the livelihood problems of communities in Jiuzhaigou, which is also the practical need of sustainable livelihood development of tourism communities in Jiuzhaigou after the earthquake.

4.5 Livelihood Strategy

Scoones [38] divided livelihood strategies into four types: expansion, intensification, diversification and migration. Among them, diversified livelihood strategy is considered as an important way of maintaining livelihood level, constructing diversified livelihood activities and realizing livelihood sustainability [37], which is conducive to reducing dependence on external environment, enhancing livelihood assets and capabilities, coping with external risk shocks and improving risk resistance. The evolution of the livelihood strategy of tourism community in Jiuzhaigou has experienced a long process. Before the development of tourism, the residents of community inside the valley adopted the livelihood strategy of agriculture. In 1987, tourism gradually developed. At that time, residents' livelihood strategy was the combination of agriculture and tourism, with main source of income from agriculture. After 1992, the livelihood strategy of community residents gradually changed to tourism. Until the earthquake, the livelihood strategies of Jiuzhaigou tourism community residents were almost all related to tourism except a few migrant workers. However, the earthquake in 2017 directly led to the interruption of tourism, which reflects the disadvantages of the single and dependent livelihood strategies of tourism communities in Jiuzhaigou. Therefore, it is necessary to choose diversified livelihood strategies for sustainable development of post-earthquake tourism communities. While adhering to the development and enrichment of tourism-related livelihood strategies, the development of non-tourism-related livelihood strategies should be encouraged, such as rental of houses, export of ethnic resources, and online sales of local specialties.

4.6 Livelihood Outcomes

Livelihood outcomes serve as standards to test whether the livelihood strategy is scientific and effective. As outcome of the livelihood strategy, it reflects the sustainability of livelihood, which is helpful to understand and improve the livelihood strategy. Before the earthquake, the tourism-related livelihood strategies adopted in Jiuzhaigou brought the corresponding livelihood outcomes to the community residents, which were composed of "increased income", "increased employment", "improved life happiness" and "spillover effects of tourism development".

At the material level, residents generally recognize the role of tourism development in increasing family income and employment, as well as improving the quality of family life. On the mental level, community residents feel an increased sense of local pride and a harmonious atmosphere of family unity. However, due to the simple livelihood strategy and external factors, the gap between the rich and the poor in Jiuzhaigou community tends to increase. At the early stage of tourism development, the income of residents inside the valley is much higher than that outside the valley, and even higher than the national level of the same period. However, the implementation of the policy of "travel in the valley, live outside the valley" leads to the relocation of business activities in the valley, and the increase of income outside the valley, leading to the dissatisfaction of residents inside the valley.

The phenomenon of spillovers of tourism development benefits started be observed in Jiuzhaigou. That is, non-local enterprises share the benefits of tourism, and the migrant population occupies the employment opportunities. The main reason for this phenomenon is that local communities lack the corresponding capital and resources to cope with the flow of tourists, thus creating opportunities for foreign enterprises and people to intercept the flow. Therefore, in the analysis of sustainable livelihood of post-earthquake tourism communities, the evaluation of livelihood outcomes can timely reflect the effect of livelihood strategies and their impact on the current livelihood capital, external environment and policies, which would effectively help community residents adjust livelihood strategies to achieve sustainable livelihood.

"C04: Before the business was ok, but then, we had a lot of people from outside, doing business in Jiuzhaigou, open restaurants, B&B, a lot of outside enterprises. Those people earn so much money, we can earn very little."

5 Discussions

While DFID's sustainable livelihood framework was proposed to address poverty alleviation and rural development, the sustainable livelihood model of Jiuzhaigou Tibetan community was proposed in the context of natural disaster, focusing on the livelihood of post-disaster tourism community at a world natural heritage site. Therefore, the development trend, seasonality and other influencing factors of tourism are proposed in livelihood vulnerability of the model. The tourism community in Jiuzhaigou has a single livelihood strategy, which is excessively dependent on tourism and lacks sustainable guarantee. Before the earthquake, the policy of "travel inside the valley, live outside the valley" was implemented, which moved out the business activities inside the valley, and the income composition of residents in the valley changed into transferred income from tickets, salary income, property income and so on, which intensified the dependence on tourism and increased the livelihood vulnerability of the tourism community in the valley. Moreover, the differences in geographical location, resources and environment of the three communities in the valley lead to the imbalance of income among the communities, which results in the dissatisfaction of residents, affects the stability and unity of the community, and further aggravates the unsustainable livelihood. Furthermore, Jiuzhaigou always attract tourists with its unique natural resources, which are easily affected by disasters and are not conducive to the sustainable development of tourism communities in the valley. Therefore, after the earthquake when the tourism development of Jiuzhaigou stopped, the residents of the tourism communities in Jiuzhaigou are facing a huge livelihood problem.

One major difference between DFID's sustainable livelihood framework and that of Jiuzhaigou is that tourism communities inside it have strong ethnic characteristics, and cultural capital is an important part of the livelihood capital. The livelihood capital of sustainable livelihood generally consists of economic, human, social, environmental and physical capital. However, some scholars gradually realize the important role of culture in sustainable livelihood [4, 43]. Traditional culture is a spiritual core deeply rooted in the hearts of residents, which always occupies an important position in the transformation of community. Local knowledge in cultural capital endows community residents with the "ability" to make a living [55], and also maintains, promotes and influences community operation and social relations, thus bringing the accumulation of social capital. Traditional culture can not only be transformed into livelihood capital and generate economic value, but also alleviate the psychological discomfort of community residents caused by livelihood transformation and tourism involvement [56]. As a world natural heritage site, Jiuzhaigou is one of the few places where residents still live inside the site, which reflects the full respect for ethnic minority residents and the protection of the relationship between people and the land. For the Tibetan communities, Jiuzhaigou is not only a place where ethnic minorities live together, but also a place where Ben-cult culture is concentrated. Ethnic culture and religious belief penetrate into every aspect of residents' life and

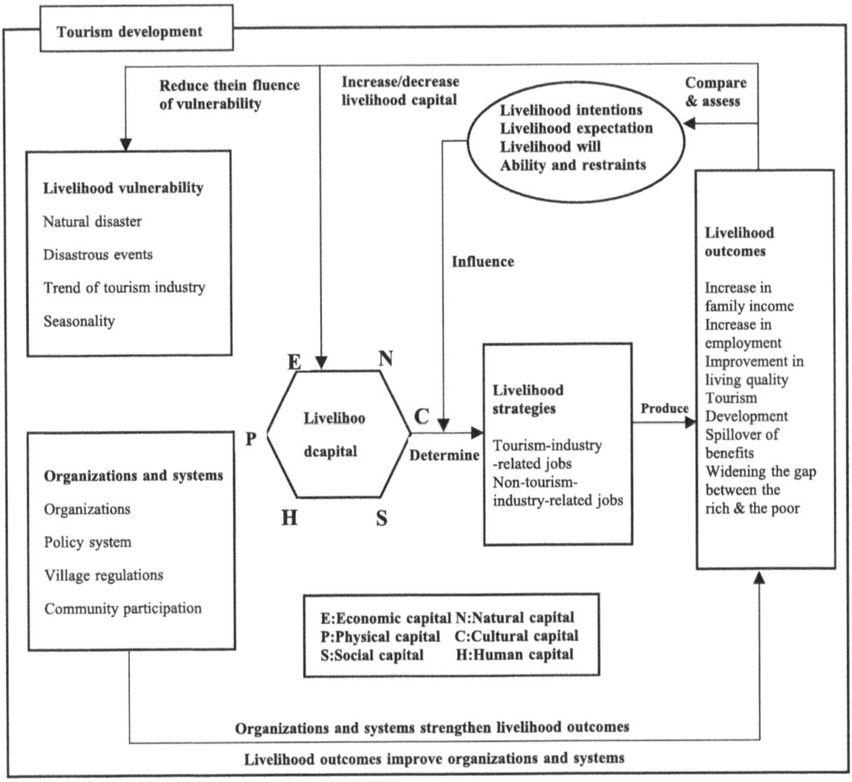

Fig. 4 Sustainable livelihood framework of post-disaster tourism community in Jiuzhaigo

work, and jointly affect and regulate the livelihood activities of residents inside the valley (Fig. 4).

During the field trip, it is found that there are a large number of cultural resources inside Jiuzhaigou valley that are not found outside, and it may serve as the livelihood capital to enrich the livelihood strategies for tourism communities inside the valley. First, the traditional Tibetan village culture. Panya, Jianpan and Yana Village are located in remote places, far away from the main scenic spots. The traditional landscape of the old village and Tibetan dwellings are still preserved, which should be protected and repaired according to the idea of "repairing the old as the old". Tourism products themed with Tibetan village culture in Jiuzhaigou can be developed to realize the activation protection of village culture and meet village residents' demand for sustainable livelihood. Second, religious belief and sacred mountain culture. Zharu Temple is the largest Ben temple near Jiuzhaigou and the only place for residents to have religious activities inside Jiuzhaigou. Zaiyizaga Sacred Mountain has been known as the "Lord of the Mountains". Before the earthquake, only local villagers, scientific research teams and a few outdoor explorers went there. Based on religious activities such as walking around the mountain and worshiping, tourism development of Jiuzhaigou after the earthquake may develop the tourism products with religious themes, so as to enrich the tourism types of Jiuzhaigou. By activating cultural capital, creating multi-type and multi-format tourism products, and livelihood strategies can be diversified.

The livelihood intention is another important factor within the post-disaster sustainable livelihood framework of tourism communities in Jiuzhaigou. Livelihood intention may serve as endogenous motivation that has moderating effect. Behavioral intention is influenced by attitude, subjective norms and perceived behavioral control [57], which directly affects human behavior. DFID's SLF covers multiple objective factors, such as livelihood capital, livelihood vulnerability and institutions, but ignores the subjective cognitive factors of community residents. Differences in individual attitudes, values, cognition and abilities often affect the choice of livelihood strategies. Based on the grounded analysis, this study found that the endogenous dynamic factor of livelihood intention plays a moderating role in the choice of livelihood strategy for post-disaster tourism communities in the valley. In the face of disasters, residents in the valley showed unity and mutual assistance, positive and enterprising national spirit and religious belief of reverence for nature, which generated the desire of sustainable development of tourism after the earthquake and the expectation to promote the sustainable development of heritage tourism through sustainable livelihood. After the earthquake, community residents have a positive livelihood intention to promote economic growth and improve the quality of life by tourism, which helps to form livelihood strategies and produce livelihood outcomes.

Community participation plays an important role in sustainable livelihoods of tourism communities. The certification of "World Heritage" gives the community capital to develop tourism and changes the development trajectory of the community. Both the development of heritage tourism and the development of the community are based on protection and inheritance. With the development of heritage tourism, communities can get better benefits from tourism. The acquisition of tourism income will strengthen community residents' awareness of heritage protection, make the heritage sustainable inheritance, and promote the sustainable development of heritage tourism community. At present, the tourism community of Jiuzhaigou World Natural Heritage Site is still in the mode of low level economic participation, and community participation is mainly in the form of symbolic community consultation/participation and material incentive participation. Jiuzhaigou has multiple management, and the division of rights and responsibilities of different management subjects is not clear, which makes it difficult to solve the problems of talent recruitment, employment, land acquisition subsidy, homestead and so on, and seriously affects the development of livelihood capital of the community. The lack of unimpeded information communication channels between the administration and residents has led to an increase in distrust between residents and the management, seriously hindered the development and mutual cooperation, and affected the recovery and sustainable development of tourism livelihoods in the valley after the earthquake. Therefore, delineating the power, responsibility and interest of community residents, establishing a relatively fair and reasonable distribution of interest coordination mechanism, satisfying needs of community residents for livelihoods, allowing community residents to express opinions, and putting forward the appeal of the platform, may help create an excellent living environment, increase livelihood capital, and provide powerful guarantee for sustainable livelihoods.

6 Conclusions

Based on DFID's SLF, this study took a post-positivist grounded theory approach to construct the sustainable livelihood model of post-disaster tourism communities at Jiuzhaigou World Natural Heritage site. It contributes to the existing study on sustainable livelihood by applying it to the context of post-disaster tourism community and further highlights the importance of cultural capital as one key component of the sustainable livelihood framework. While the post-earthquake period is the most vulnerable period for tourism community in Jiuzhaigou, it is also the key period for the development of the Tibetan community. The unique geographical environment has created the outstanding natural landscape of Jiuzhaigou World Natural Heritage. Residents of the community benefit from the development of tourism. Meanwhile, Jiuzhaigou is located in a zone of frequent natural disasters, which can damage the livelihoods of the community that depend entirely on tourism. In order to reduce the vulnerability and risk of livelihood, the upgrading of tourism industry and diversification of livelihood are important for the sustainable livelihood of tourism community in Jiuzhaigou. In fact, Jiuzhaigou not only has unique natural resources, but also has rich cultural resources that are not found outside the valley. To introduce the original Anduo Tibetan ethnic culture into the tourism industry, the livelihood capital of local communities would be more diverse. At the same time, more attention should be paid to residents' intentions, stimulating residents' subjective initiative, so as to realize the harmony between man and nature, and ensure the comprehensive coordination and sustainable development of community industry.

As data about the communities inside the valley is difficult to get, this study contributes to the existing study on tourism communities at Jiuzhaigou World Natural Heritage site. As a typical community living in the World Heritage site, the transformation of community livelihood of Jiuzhaigou can provide reference for the livelihood research on communities at other World Heritage Site. In addition, with the ongoing outbreak of natural disasters and crises such as the coronavirus pandemic, the livelihood development of tourism communities and post-disaster reconstruction will continue to attract worldwide attention in the long run. Further study can be done to quantify the influencing factors found in the qualitative research, and analyze the effects of sustainable livelihood of tourism communities after the earthquake through quantitative research, so as to obtain more scientific and reliable research conclusions.

References

1. Pforr, C., Hosie, P.J.: Crisis management in tourism—preparing for recovery. J. Travel Tour. Mark. **23**(2), 249–264 (2008)
2. Mair, J., Ritchie, B.W., Walters, G.: Towards a research agenda for post-disaster and post-crisis recovery strategies for tourist destinations: a narrative review. Curr. Issue Tour. **19**(1), 1–26 (2016)
3. Edwards, P., Sharma-Wallace, L, Barnard, T. et al.: Sustainable livelihoods approaches to inform government-local partnerships and decision-making in vulnerable environments. New Zealand Geographer 75, 63–73 (2019)
4. Tao, T., Wall, G.: A livelihood approach to sustainability. Asia Pacific Journal of Tourism Research **14**(2), 137–152 (2009)
5. Wu, M., Pearce, P.L.: Host tourism aspirations as a point of departure for the sustainable livelihoods approach. J. Sustain. Tour. **22**(3), 440–460 (2014)
6. Su, M. M., Sun, Y. H., Jiao, W. J., Min, Q. W.: A community livelihood approach to agricultural heritage system conservation and tourism development: Xuanhua Grape Garden urban agricultural heritage site, Hebei Province of China. Sustainability 10(2), 361 (2018). https://doi.org/10.3390/su10020361
7. Job, H., Becken, S., Lane, B.: Protected Areas in a neoliberal world and the role of tourism in supporting conservation and sustainable development: an assessment of strategic planning, zoning, impact monitoring, and tourism management at natural World Heritage Sites. J. Sustain. Tour. **25**(12), 1697–1718 (2017)
8. Millar, S.: Heritage management for heritage tourism. Tour. Manage. **10**(1), 9–14 (1989)
9. Yuksel, F., Bramwell, B., Yuksel, A.: Stakeholder interviews and tourism planning at Pamukkale. Turkey. Tourism Manag. **20**, 351–360 (1999)
10. Araújo, L.: Can tourism value heritage? Worldwide Hospitality and Tourism Themes **11**(6), 692–695 (2019)
11. Shen, F., Hughey, K.F.D., Simmons, D.G.: Connecting the sustainable livelihoods approach and tourism: a review of the literature. J. Hosp. Tour. Manag. **15**(1), 19–31 (2008)
12. Tao, T., Wall, G.: Tourism as a sustainable livelihood strategy. Tour. Manage. **30**(1), 90–98 (2009)
13. Stone, M.T., Nyaupane, G.P.: Protected areas, wildlife-based community tourism and community livelihoods dynamics: Spiraling up and down of community capitals. J. Sustain. Tour. **26**(2), 307–324 (2018)
14. Adiyia, B., Vanneste, D., Van Rompaey, A.: The poverty alleviation potential of tourism employment as an off-farm activity on the local livelihoods surrounding Kibale National Park, western Uganda. Tour. Hosp. Res. **17**(1), 34–51 (2017)
15. Matiku, S., Zuwarimwe, J., Tshipala, N.: Community-driven tourism projects' economic contribution to community livelihoods—A case of Makuleke contractual park community tourism project. Sustainability 12, 8230 (2020). https://doi.org/10.3390/su12198230
16. Nepal, S.K.: Sustainable tourism, protected areas and livelihood needs of local communities in developing countries. Int. J. Sust. Dev. World **4**, 123–135 (1997)
17. Sirima, A., Backman, K.F.: Communities' displacement from national park and tourism development in the Usangu Plains, Tanzania. Curr. Issues Tourism **16**(7–8), 719–735 (2013)
18. Su, M.M., Wall, G., Xu, K.J.: Heritage tourism and livelihood sustainability of a resettled rural community: Mount Sanqingshan World Heritage Site, China. J. Sustain. Tourism **24**(5), 735–757 (2016)
19. Lasso, A., Dahles, H.: Are tourism livelihoods sustainable? Tourism development and economic transformation on Komodo Island, Indonesia. Asia Pacific J. Tourism Res. **23**(5), 473–485 (2018)
20. Melubo, K., Lovelock, B.: Living inside a UNESCO World heritage site: the perspective of the Maasai community in Tanzania. Tourism Plan. Develop. **16**(2), 197–216 (2019)
21. Weir, D.: Restoring tourism destinations in crisis: a strategic marketing approach. Tour. Manage. **26**(3), 475–476 (2005)

22. Rittichainuwat, B.N.: Tourists' and tourism suppliers' perceptions toward crisis management on tsunami. Tour. Manage. **34**, 112–121 (2013)
23. Gawith, E.: The on-going psychological toll from the Canterbury earthquakes: Stories from one community. Disaster Prev. Manag. **22**(5), 395–404 (2013)
24. Tsao, C.Y., Ni, C.: Vulnerability, resilience, and the adaptive cycle in a crisis-prone tourism community. Tour. Geogr. **18**(1), 80–105 (2016)
25. Tsai, C., Chen, C.: An earthquake disaster management mechanism based on risk assessment information for the tourism industry. Tour. Manage. **31**, 470–481 (2010)
26. Lee, W., Chhabra, D.: Heritage hotels and historic lodging: perspectives on experiential marketing and sustainable culture. J. Herit. Tour. **10**(2), 103–110 (2015)
27. Nian, S.F., Zhang, J., Zhang, H.L., et al.: Two sides of a coin: A crisis response perspective on tourist community participation in a post-disaster environment. Int. J. Environ. Res. Public Health **16**(12), 2073 (2019)
28. Cope, M.R., Slack, T., Blanchard, T.C., et al.: Does time heal all wounds? Community attachment, natural resource employment, and health impacts in the wake of the BP Deepwater Horizon disaster. Soc. Sci. Res. **42**(3), 872–881 (2013)
29. Vallance, S., Carlton, S.: First to respond, last to leave: Communities' roles and resilience across the '4Rs.' Int. J. Disaster Risk Reduction **14**, 27–36 (2015)
30. Parsons, M., Glavac, S., Hastings, P., et al.: Top-down assessment of disaster resilience: A conceptual framework using coping and adaptive capacities. Int. J. Disaster Risk Reduction **19**, 1–11 (2016)
31. Bachrach, K.M., Zautra, A.J.: Coping with a community stressor: the threat of a hazardous waste facility. J. Health Soc. Behav. **26**(2), 127–141 (1985)
32. Kim, K., Uysal, M., Sirgy, M.J.: How does tourism in a community impact the quality of life of community residents? Tour. Manage. **36**, 527–540 (2013)
33. Lin, Y.W., Kelemen, M., Tresidder, R.: Post-disaster tourism: building resilience through community-led approaches in the aftermath of the 2011 disasters in Japan. J. Sustain. Tour. **26**(10), 1766–1783 (2018)
34. Goulding, C., Kelemen, M., Kiyomiya, T.: Community based response to the Japanese Tsunami: a bottom-up approach. Eur. J. Oper. Res. **268**(3), 887–903 (2018)
35. Kato, K.: Debating sustainability in tourism development: resilience, traditional knowledge and community: a post-disaster perspective. Tourism Plan. Develop. **15**(1), 55–67 (2018)
36. Chambers, R., Conway, G.: Sustainable rural livelihoods: practical concepts for the 21st century. Institute of Development Studies (UK) (1992)
37. Ellis, F.: Rural livelihoods and diversity in developing countries. Oxford University Press, Oxford (2000)
38. Scoones, I.: Sustainable rural livelihoods: a framework for analysis. University of Sussex, Brighton (1998)
39. Mbaiwa, J.E., Stronza, A.L.: The effects of tourism development on rural livelihoods in the Okavango Delta, Botswana. J. Sustain. Tourism **18**(5), 35–56 (2010)
40. Snider, R.: Land tenure, ecotourism, and sustainable livelihoods: "Living on the Edge" of the Greater Maasai Mara, Kenya. University of Waterloo, Canada (2012)
41. DFID.: Sustainable Livelihoods Guidance Sheets. Department for International Development, London (2000)
42. Ashley, C., Carney, D.: Sustainable livelihoods: Lessons from early experience. Department for International Development, London (1999)
43. Su, M.M., Wall, G., Jin, M.: Island livelihoods: tourism and fishing at Long Islands, Shandong Province, China. Ocean Coast. Manage. **122**, 20–29 (2016)
44. Su, M.M., Wall, G., Wang, Y., Jin, M.: Livelihood sustainability in a rural tourism destination—Hetu Town, Anhui Province, China. Tourism Manage. **71**, 272–281 (2019)
45. Srijuntrapun, P., Fisher, D., Rennie, H.G.: Assessing the sustainability of tourism-related livelihoods in an urban World Heritage Site. J. Herit. Tour. **13**(5), 395–410 (2018)
46. Aazami, F., Shanazi, K.: Tourism wetlands and rural sustainable livelihood: the case from Iran. J. Outdoor Recreat. Tourism 30, 100284 (2020). https://doi.org/10.1016/j.jort.2020.100284

47. Glaser, B., Strauss, A.: The discovery of grounded theory: strategies for qualitative research. Aldine, Chicago (1967)
48. Strauss, A., Corbin, J.: Basics of qualitative research: techniques and procedures for developing grounded theory. Sage, CA (1998)
49. Charmaz, K.: Constructing grounded theory: a practical guide through qualitative analysis. Sage, London (2006)
50. Bryant, A., Charmaz, K.: Introduction: grounded theory research: methods and practices. In: Bryant, A., Charmaz, K. (eds.) The Sage Handbook of Grounded Theory, pp. 1–28. Sage, London (2007)
51. Charmaz, K.: Grounded theory as an emergent method. In: Hesse-Niber, S.N., Leavy, P. (eds.) Handbook of Emergent Methods, pp. 155–172. The Guilford Press, New York (2008)
52. Charmaz, K.: Grounded theory methods in social justice research. In: Denzin, N.K., Lincoln, Y.S. (eds.) The 4th Sage Handbook of Qualitative Research, pp. 359–380. Sage, Los Angeles (2011)
53. Matteucci, X., Gnoth, J.: Elaborating on grounded theory in tourism research. Ann. Tour. Res. **65**, 49–59 (2017)
54. Sang, Z.: Tourism development and spatial culture reconstruction of Jiuzhaigou. J. Southwest Minzu University (Humanities and Social Sciences Edition) **5**, 47–50 (2014)
55. Delaney, A.: Social sustainability in post-3.11 coastal Japan: The significance of social capital [M]. In Sustainability in contemporary rural Japan: Challenges and opportunities. Ins. Assmann ed., New York, 3–18 (2016)
56. Daskon, C.D.: Cultural resilience—the roles of cultural traditions in sustaining rural livelihoods: a case study from rural Kandyan villages in Central Sri Lanka. Sustainability **2**(4), 1080–1100 (2010)
57. Ajzen, I.: Perceived behavioral control, self-sufficacy, locus of control, and the theory of planned behavior. J. Appl. Soc. Psychol. **32**(4), 665–683 (2002)

Forty-Seven Years of Museum Tourism Research: A Bibliometric and Network Analysis

Yanfei Huang, Xiaowen Hu, Wei Ma, and NaZhaKaiTi Yusupu

Abstract Museum tourism is of great significance in the protection of cultural heritage. The purpose of this study is to systematically analyze the literature on museum tourism research through bibliometrics. For this purpose, 161 papers were retrieved from the Web of Science database. Qualitatively and quantitatively study is performed, and four major themes are identified. The main theme in the papers is concerned with the experience of museum tourism. After discussing the regional, contemporary, and inclusive characteristics of the hot topics in the museum tourism research, we discover that the research literature reflects the transformation from personal feelings to public practice. This highlights the significance of the museum tourism for the national consciousness and national values. Based on the confirmed findings, the paper recommends several directions for future research in this area.

Keywords Museum tourism · Systematic review · Bibliometric analysis · CiteSpace

Y. Huang · X. Hu (✉) · W. Ma · N. Yusupu
Northwest Minzu University, Lanzhou, Gansu, People's Republic of China
e-mail: 296079144@qq.com

Y. Huang
e-mail: 1711130375@qq.com

W. Ma
e-mail: 306363677@qq.com

N. Yusupu
e-mail: 2176264389@qq.com

X. Hu
Sichuan University, Chengdu, Sichuan, People's Republic of China

Y. Luo et al. (eds.), *Tourist Experience and Perception in the Pandemic and Post-Pandemic Era*, https://doi.org/10.1007/978-981-96-1820-0_10

1 Introduction

Human civilization needs to be protected. With the progress of the human economy and civilization, the number of museums, exhibition themes, and tour forms has been enriched. This has attracted many scholars to study the field of museums, while also meeting people's spiritual and cultural needs.

As we all know, the concept of a museum has not been accurately defined and has been discussed by scholars in academic circles. Günay [1] discusses the source of the concept of the museum and believes that it is a social platform that combines modern educational organizations. Zou [2] argues that the museum, as a physical space, serves as a system for recording historical events and is also a cultural symbol. Tigert [3] suggests that the museum is a unique space for social learning, providing people with an informal learning environment. The 26th International Association of Museums, held in Prague, adopted a new definition of a museum, which defines it as a non-profit permanent institution serving society by showcasing various meaningful materials and intangible heritages, and providing people with diverse cultural experiences. Since 1946, the International Council of Museums has continuously revised the definition of museums. It can be seen that the definition of museums keeps pace with the times and also reflects the continuous concern of society for museums.

Lin [4] examined how museums serve different forms of leisure, including casual leisure, project-based leisure, and serious leisure. Serious leisure involves the dedicated and systematic pursuit of activities such as amateur interests, hobbies, or volunteer work [5]. It can be seen that the museum plays a cognitive and educational role in science, art, culture, and entertainment. When conducting amateur activities, hobbyists meet the purposeful needs of hobbyists in collecting materials, repairing museums, and participating in museum activities. The volunteer core activity is a way for volunteers to realize their needs of self-realization through their altruism. Therefore, museums are essential places for serious leisure. Cultural tourism can be regarded as a form of serious leisure [6], and museums are a part of cultural tourism. Combined with the particularity of museums, it can be considered that museums are indispensable and important places for serious leisure. In the future, museums may be an important field for serious leisure research.

As an important place for informal learning, museums have a positive impact on the audience's cognitive development, emotional attitude, and social skills. Museums began to focus on people as the center and opened up a new field of museum pedagogy, making people realize that museums are not only the property of rulers and the bourgeoisie [1]. When disaster prevention education activities are carried out in museums, students' awareness of disaster prevention learning significantly improves [7]. Heritage tourism, specifically through museums, can help rebuild national identity, national consciousness, and a sense of belonging [8, 9]. Distance learning in museums has improved learners' learning efficiency and communication skills [10]. In museums, learners have the opportunity to meet people from different backgrounds and cultures, which helps improve their social communication skills. Additionally, the informal learning environment of the museum encourages the cultivation of self-expression and creativity through participation in various practical activities.

However, despite the increasing attention paid to research on museum tourism, there is still a lack of literature reviews on the overall situation of this field. Most of the research still focuses on analyzing tourists' experiences and perceived reasons. Using bibliometrics, this paper quantitatively discusses the literature-based situation of museum tourism research, including its main hot topics.

2 Research Methods and Data Sources

Bibliometric is a quantitative study of the substitution of publishing units and bibliographic units, or both [11]. Bibliometric analysis focuses on database documents as its research object and utilizes econometric research methods to study the distribution structure, quantitative relationships, and changing patterns of document information [12]. Bibliometric has numerous sources of literature data, which can yield different statistical results depending on how they are processed [13]. It can be seen that the source of data in bibliometric analysis is very strict.

CiteSpace is a Java-based tool used to model and visualize networks, with the primary objective of facilitating the exploration of evolving patterns in academic literature [14]. This study utilizes CiteSpace as a bibliometric tool to conduct co-word analysis, cluster analysis, and co-citation analysis on data related to museum tourism research. One of the key techniques applied in this analysis is cluster analysis, which can be carried out in two distinct ways: through co-cited documents and via co-occurring keywords [15]. This analysis utilizes cluster analysis in two distinct ways: through co-cited documents and via co-occurring keywords [15]. If two articles are co-cited by a third article, it is called co-citation [16]. In this paper, a keyword knowledge map is used for cluster analysis to explore scholars' research hotspots. Co-citation is used to analyze the citation frequency of scholars' articles.

WOS, which is a product of Thomson Reuters, is an extensive authoritative database that includes information dating back to 1900 [17]. It contains more than 8,000 influential high-quality journals from all over the world. Moreover, the multi-disciplinary nature of WOS, the availability of references, and other characteristics have kept it at the forefront for decades [18]. Therefore, this paper chooses the Web of Science database as the source of original data.

The retrieval time of the data in this article is September 1, 2022. To ensure the comprehensiveness of the retrieval data, the subject term of the retrieval is "museum tourism", the article type is "article", and the publication time is unlimited. After the retrieval, the above documents were reviewed, and 161 documents were obtained by manually reading the titles and abstracts, and removing irrelevant documents, repetitive documents, book reviews, advertisements, and other materials.

3 Results

3.1 Publications on Museum Tourism

3.1.1 The Number of Published Documents and Their Citation Frequency

As shown in Fig. 1, from 1975 to 2022, the number of publications in the field of museum tourism research and their citations has increased.

Overall, the number of articles being published in museum tourism research is steadily increasing each year. While there have been instances of explosive growth in the publication of articles in this field, the pace has somewhat slowed in recent years. This indicates sustained interest from scholars in this research area. Over time, the articles related to tourism in museums can be categorized into four distinct stages, each with its own unique research focal points Fig. 2.

In the first stage (1975–2004), a total of 39 articles were published in this field over the course of 29 years. Although scholars have increased their research on museum tourism, they have paid less attention to this field. In the second stage (2005–2007), four articles were published during the three-year period. Among these, scholars' research on museum tourism decreased, indicating a weakening of attention in this field. In the third stage (2008–2014), scholars' research on museum tourism increased and then decreased. At this stage, the number of articles published was 22. From 1998 to 2012, the United States studied national buzzwords and paid the highest attention to this field. In the fourth stage (2015–2022), the number of articles published was the largest, with 96 articles. Among them, from 2015 to 2022, scholars' research in the field of museum tourism maintained steady growth. This shows that this field has begun to attract attention, and related research is becoming more popular. This change is in line with the political and economic environment of the times, and some countries and institutions are promoting the development of museum tourism research.

3.1.2 Research Hotspots and Clustering of Museum Tourism

Keywords reveal different themes and confirm interest in this theme [19]. Clustering technology is used to study and analyze the main subject or research field of a certain scientific field [20]. Using CiteSpace software is used to conduct the co-occurrence spectrum analysis of keywords in museum tourism research literature from 1975 to 2022 (as shown in Fig. 3). The key word with a high frequency in museum tourism research is "experience", "cultural tourism", "satisfaction", "model", "dark tourism", "motivation" etc. The results of refining high-frequency keywords approximately reflect the research theme, essential content, and research methods of scholars in the area of museum tourism, which are generally consistent with the major policies

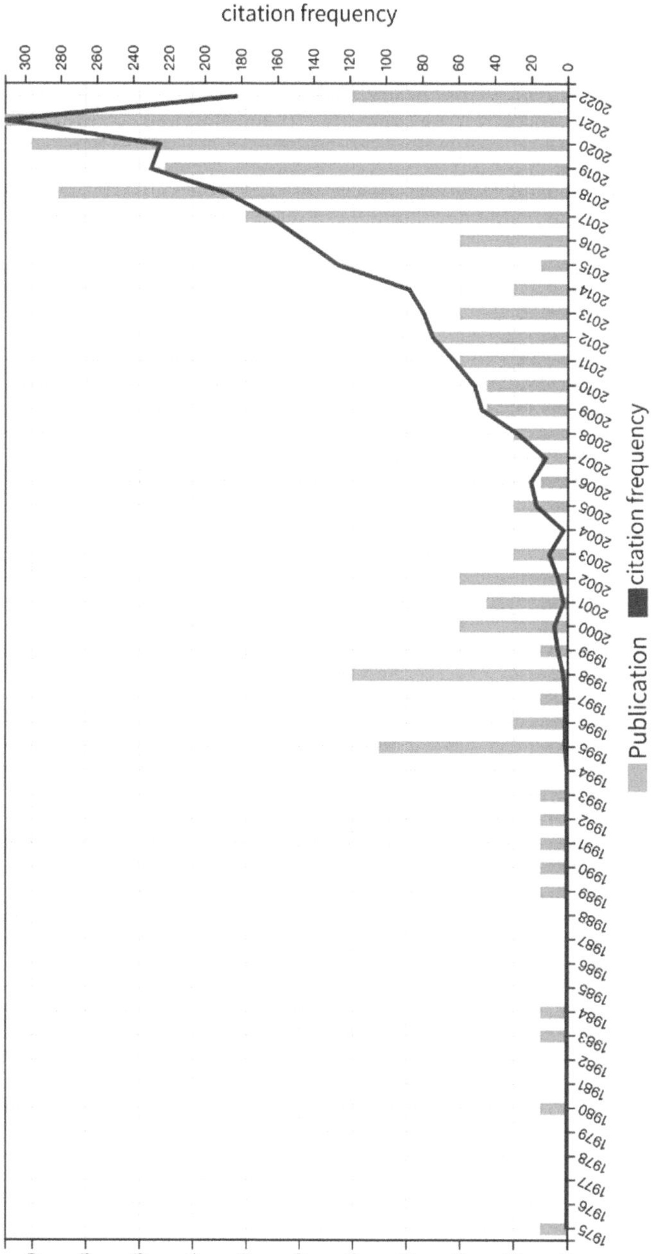

Fig. 1 The time evolution of the number of publications and the cited frequency of museum tourism research

proposed by various countries, the development of related industries, and social needs.

By using the LLR algorithm in CiteSpace software, this paper presents a cluster analysis of the co-occurrence keywords of museum tourism research, identifying 10 cluster themes. The lack of flexibility in software clustering makes it difficult to conduct a clustering analysis. Consequently, 10 museum tourism research topics were further evaluated and classified according to classification. The hot topics of museum tourism research are summarized as a variety of museum types, the museum tourism experience, sustainable development of museums, and the museum tourism effects (Table 1).

Table 1 Cited authors of museum tourism research literature

Rank	Author	Frequency	Year
	Richards G	13	1998
	Packer J	10	2017
	Falk JH	10	2019
	Falk J	10	1996
	Mckercher B	9	2016
	Cohen E	9	1991
	Poria Y	9	2016
	Sheng CW	8	2018
	Su YG	7	2019
	Plaza B	6	2000
	Stebbins RA	6	2019
	Jansenverbeke M	6	1996
	Ashworth GJ	5	2011
	Parasuraman A	5	2020
	Anton C	5	2019
	Kim S	5	2020
	Buhalis D	5	2020
	Ashworth G	4	1996
	Bennett TONY	4	1997
	Lennon J	4	2011
	Urry J	4	2017
	Mcpherson G	4	2019
	Moscardo G	4	1991
	Chan JKL	4	2020
	Ryan C	4	2016

3.1.3 The Author's Contribution to Museum Tourism Research

The articles frequently cited by scholars reflect the frontier research in this field to some extent [21]. In order to better explore the in-depth research of scholars on museum tourism and the output rate of achievements in this field, the frequency of the top 25 cited authors and co-authors was analyzed. Additionally, a co-citation analysis of literature related to museum tourism research was conducted, using keyword clustering as a basis. The articles by Kim S, Plaza B, and Park E are highly cited articles, and most of them are cited in 2020(as shown in Fig. 4). Su YG, Anton C, and Bideci M are scholars who frequently co-author with other scholars and who are frequently cited. Most of the results are concentrated in 2018, and the research focuses on the tourist experience [22–24] (as shown in Tables 2 and 3). Although most of these authors' achievements in the field of museum tourism research have been concentrated in recent years, they have had a far-reaching impact.

It has been found that Belgium and the USA are the most influential countries in early museum tourism research. It was not until 2010 that Asian countries, represented by China and Singapore, attracted attention in this field. Among them, the research institutions with the greatest influence are represented by Hong Kong Polytechnic University and City University of Macau in Asia.

3.1.4 The Influence of Top Journals by Publication

The influence of published journals related to museum tourism research is mainly determined by the citation frequency of articles published in this field. The higher the citation frequency of published journals, the greater the interest and attention of journals in this field, and the greater their contribution to the future development and academic research in this field [16].

The top 10 most frequently cited journals (as indicated in Table 3) reveal that the Annals of Tourism Research is the foremost influential journal in the field of museum tourism, followed by Tourism Management and the Journal of Travel Research. The Annals of Tourism Research serves as a crucial platform for exploring research on museum tourism, with its articles being cited frequently and early on, thereby contributing significantly to further research in this area. Additionally, Tourism Management, as a secondary source for studying museum tourism research, has a high frequency of citations and early publication dates, playing a pivotal role in advancing this growing field. Moreover, the Journal of Travel Research holds the third position in citations related to museum tourism research, with citations increasing over time.

Table 2 Cooperative authors in museum tourism research

Rank	Author	Frequency	Year
	Su YG	7	2018
	Anton C	4	2018
	Bideci M	3	2018
	Falk J	2	1992
	Jansenverbeke M	2	1996
	Plaza B	2	1999
	Krens T	2	1999
	Stone P	2	2008
	Cooke P	2	2008
	Chhabra D	2	2008
	Cherif R	2	2009
	Yang CH	2	2010
	Hall CM	2	2011
	Larsen J	2	2011
	Isaac RK	2	2014
	Hartmann R	2	2014
	Waterton E	2	2014
	Moscardo G	2	2015
	Kim S	2	2015
	Di PIETROL	2	2015
	Carter PL	2	2016
	Falk J	2	2016
	Kim S	2	2016
	Wu KY	2	2016
	Gretzel U	2	2016

Table 3 Cited journals in museum tourism research

Rank	Source	Frequency	Year
	Annals of Tourism Research	68	1991
	Tourism Management	50	2000
	Journal of Travel Research	29	2016
	Journal of Travel Research	23	2002
	International Journal of Heritage Studies	22	2006
	Current Issues in Tourism	20	2019
	Journal of Travel & Tourism Marketing	19	2015
	International Journal of Tourism Research	19	2019
	Tourism Management Perspectives	18	2019
	Museum Management and Curatorship	17	2015

3.2 Thematic Analysis: Subtopics of Museum Tourism Research

3.2.1 Types of Museums

All types of tourism studies suggest that various types of museums, including stamp collections, historical institutions, and children's museums, are unique cultural attractions [25]. Some museums are characterized by political color and dark tourism [26] and are studied in a narrative way, which is consistent with the analysis in Fig. 2. Plant museums reflect the American Civil War [27]. Andermann [28]argued about the concept of 'space for memory' from a narrative perspective. The well-established museum, sponsored by enterprises, showcases Beijing Roast Duck and highlights the current dynamics of the relationship between the state and society in China [29].

Most museums use case studies to analyze the current situation or challenges of museums. For example, takes the Cobdogla Irrigation and Steam Museum, which is managed by volunteers, as an example to analyze the current situation and propose relevant solutions [30]. From the case analysis, the strategy of developing sports museums was proposed for sports heritage from the perspective of virtual employees [31]. Children's museums can be summarized by the challenges and problems they face. [32]. There are also related studies on other types of museums, such as assemblage museums and archaeological museums [25, 33]. There are many kinds of museums, which are accompanied by cultural factors in tourism [34]. It has played a significant role in the construction of global culture. The strong inclusiveness of museums enriches the types of museums, and also meets people's demand for knowledge of Maslow's hierarchy of needs in the spiritual world after the epidemic.

3.2.2 Museum Tourism Experience

For over 30 years, the study of tourism has become one of the most popular academic subjects [35]. Researchers have extensively studied the experiences of museum visitors, making it the most researched aspect of museum tourism research. Certain academics have started employing content analysis techniques to examine museum reviews found on online travel platforms [36]. However, there are few applications that focus on museum travel experiences, such as extracting the dimensions of service quality from negative online reviews and emphasizing the impact of museum service failures on customer satisfaction and experience [24].

Many scholars use questionnaires to measure the variables in the model. Antón [22] identified the factors that either prevent or deepen tourists' satisfaction during their visits. Bideci and Albayrak [23] also examined the fundamental dimensions of the museum experience, including entertainment, comfort, escape, and aesthetics, to evaluate the differences in tourists' perceptions of museum experiences at home and abroad. Di Pietro [37] emphasized the crucial role of technology in shaping the tourist experience. Shao [36] determined that the cost of paid products and services,

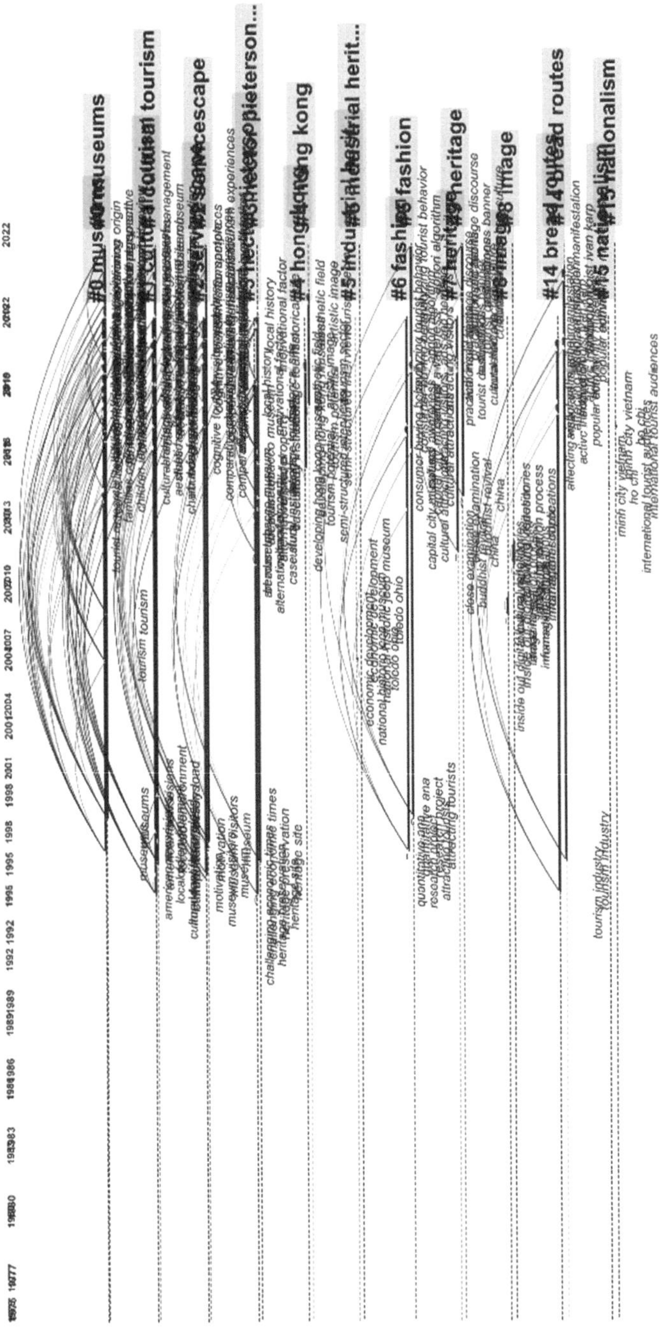

Fig. 2 Hotspots in various stages of museum tourism research

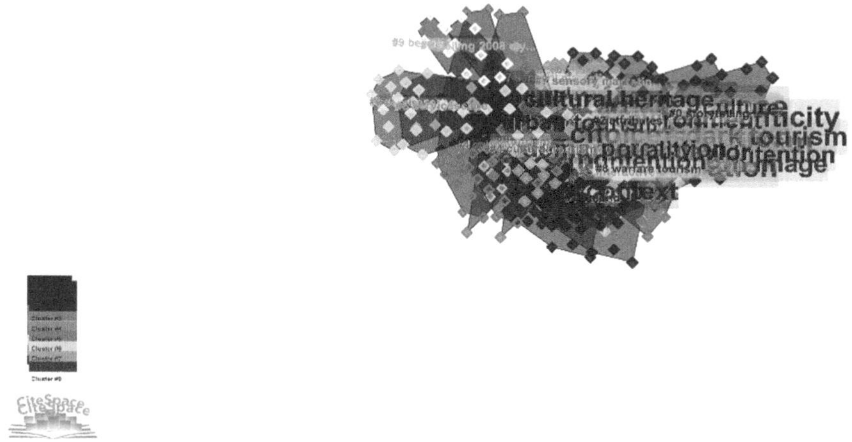

Fig. 3 Research hotspots and clustering of museum tourism

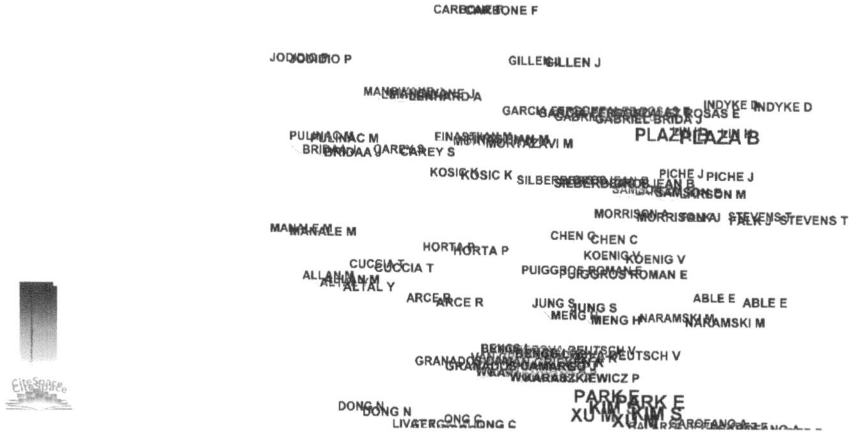

Fig. 4 Cluster atlas of cited articles on museum tourism research

the attitude of staff, and the environmental facilities of museums are crucial factors that impact the tourist experience.

Additionally, Antón [38] analyzed the factors that either prevent or deepen tourists' satisfaction. The museum tourism experience is based on tourists' physical and mental sensations, and it can also be influenced by their perception of images [39] and the overall image of the tourist destination [40].

3.2.3 Sustainable Development of Museum Tourism

In urban tourism, the development of museums has hardly been addressed [41]. Elites from all walks of life constantly adopt various policies and measures to ensure the sustainable development of museum tourism.

Some scholars explored the sustainable development of museum tourism from the perspective of public services. There are twelve proposed kinds of service quality for museums to take into account when considering cultural tourism [24]. These twelve proposed kinds of service quality take into account the causes of service failure and solutions. RIM provides many facilities and services that tourists expect from museums [34]. Effective tourist information services are crucial to the development of tourist destinations [42]. Some scholars also explored the sustainable development model of museum tourism from the perspective of world heritage. It was believed that the number of world heritage sites, regions, and tourists attracted to tourism destinations were also different [43]. A business model for the development of cultural heritage whereby income from tourism is used to protect cultural relics collections and buildings that collect cultural relics [44].

In addition, some scholars have studied the sustainable development of museums from the perspective of individual differences in museums. locating curators in the authenticity continuum, and putting forward further requirements for the authenticity development of museums [43]. ML and MXL models analyze the personal preferences of different museum services, and the resulting services aim to optimize accessibility and understanding, and improve the educational purpose of the visit [45]. The key to sustainable development in museum tourism lies in providing excellent public services, as well as understanding the value and purpose of museums.

3.2.4 Museum Tourism Effects

Undoubtedly, tourism is now a major industry in the world [43]. The elements and institutions involved in museum tourism are extremely rich, and the related industries are closely linked. For example, in the cultural industry, the museum itself is an integral part of the cultural and artistic industries, bearing and protecting the cultural heritage of mankind. In terms of tourism, museums serve as popular tourist destinations and directly contribute to the growth of the tourism industry. The information technology industry plays a vital role. Modern museums typically need to utilize advanced information technology to enhance visitors' experiences, such as digital displays and virtual reality (VR). It can be seen that museum tourism has had a far-reaching impact on the economy, politics, and education.

First of all, museum tourism can promote the economy. For example, the tourist consumption pattern of heritage tourism has been studied. The results show that the average total expenditure of tourists visiting farms, museums, parks, and gardens is the most significant [46]. Using museums as an effective marketing tool for urban tourism has a significant impact on the economic development of museum tourism [41].

Secondly, museum tourism strengthens tourists' cognition and emotional attitude. During the 2008 Beijing Olympic Games, museum tourism indirectly promoted the integration of China with society [29]. Tolia-Kelly discussed the national and political sentiments of colonial museums [47]. Through the museum experience, it is considered that museums are important places for national identity [48]. Museum tourism enhances tourists' awareness of cultural respect, cultural identity, and cultural output.

Finally, museum tourism strengthens the education of tourists, which is mainly reflected in personal and national education. In terms of personal education, museum tourism utilizes augmented reality (AR) to realize the economic value, experiential value, cognitive value, and educational value of AR [49]. Children's museums will have a significant impact on the educational effect of cognitive development [50]. Museum tourism can shape values [51]. In terms of national education, museum tourism can deepen a national cultural identity, promote patriotic education, and foster a sense of belonging to the country and the nation [52, 53].

To sum up, the influence of museum tourism is subtle, long-lasting, and extensive in scope.

4 Conclusions

Using data from the Web of Science database, this paper conducts a qualitative thematic review and a quantitative systematic review of research on museum tourism from 1975 to September 2022. It explores popular topics, methods, and development trends in museum tourism research and analyzes the shortcomings of museum tourism research. The following conclusions can be drawn:

In the research of hot topics, four hot topics were obtained through cluster analysis and manual processing. Then, according to keyword frequency analysis, the results focused on the research of museums and museum visitors. For museums, the initial research mainly focuses on the case method and puts forward the threats, challenges, and solutions faced by museums. The limitation is that some museum solutions are not always universal. For museum visitors, based on studying museums, we should examine the factors that affect visitors' experiences, perceptions, and motivations from the perspectives of psychology, economics, and sociology. We should constantly enrich the achievements of the theme of museum tourism experience. In the future, research can be carried out based on the COVID-19 epidemic and post-COVID-19 epidemic, as well as factors such as gender, race, value co-creation, and value co-destruction. This will lay the foundation for the development of thematic museums.

According to the co-cited documents, the highly cited documents mainly use empirical methods in their research. Statistical analysis involves examining samples or data, while also utilizing models and theoretical analysis from fields such as psychology, sociology, and economics. Although most literature uses questionnaires to test relevant research models, some articles are gradually using big data methods to crawl online comments for research. However, some models still require further study. Therefore, it is the only way to strengthen the rationalization, innovation, and in-depth research of related models in the future.

In terms of research trends, museums are often associated with the heritage of cities and tourist destinations, which is in line with the latest definition of museums by the International Council of Museums. More cases are being used to analyze the current situation and study potential solutions, but there is no research on the integrity of the museum community and the functional nature of museums. These aspects can be strengthened in the future. Moreover, research in the field of museum tourism is closely related to cultural tourism, dark tourism, and heritage tourism. This has led to an increase in research on museum tourism and the expansion of research areas from south to north and from west to east around the world. However, there are few studies on the integration of museums with other industries. In the future, we will focus on excavating the cultural connotation of museums. We will also pay special attention to integrating museums with technology industries such as AR, VR, and MR. Through this, we aim to continuously address the challenges faced by museums and promote the implementation of various modes.

5 Limitations

This study also has some limitations. First of all, the retrieval date for this study is September 1, 2022, and the collection of research data in this field in 2022 is incomplete. In the future, the relevant data from museum tourism research in 2022 can be further improved and analyzed. Secondly, the original data was downloaded from the SSCI database of the WOS core collection retrieval, and there are only English publications. Therefore, many publishing languages can be considered in the future. Thirdly, only WOS is used as a database. Due to the need to extract a complete list of references and citations, there is a lack of reliable automatic duplicate deletion for large datasets with incomplete digital object identifiers (DOI). Multiple databases can be used for data screening in the future [54].

Acknowledgements This work was supported by the Basic Scientific Research Project of Central University (No. 31920220119 and No. 31920240034), the Innovation and Entrepreneurship Education Reform Project in Provincial Colleges and Universities (No. 100154010705), the Northwest Minzu University Undergraduate Training Quality Improvement Project (No. 2022XJJG-19), and the Northwest Minzu University 2023 Provincial College Students' Innovation and Entrepreneurship Training Program Project (No. S202310742002O).

References

1. Günay, B.: Museum concept from past to present and importance of museums as centers of art education. Proc. Soc. Behav. Sci. **55**, 1250–1258 (2012)

2. Zou, Y., Xiao, H., Yang, Y.: Constructing identity in space and place: semiotic and discourse analyses of museum tourism. Tour. Manage. **93**, 104608 (2022)
3. Tigert, J.M., Fotouhi, G., Kirschbaum, S.: An investigation of museum educators' questioning during field trips. Learning, Culture and Social Interaction 31, 100571 (2021)
4. Lin, Y.: Leisure—a function of museums? The Taiwan perspective. Museum Manage. Curatorship **21**(4), 302–316 (2006)
5. Stebbins, R.A.: Leisure and its relationship to library and: information science: bridging the gap. Libr. Trends **57**(4), 618–631 (2009)
6. Stebbins, R.A.: Cultural tourism as serious leisure. Ann. Tour. Res. **23**(4), 948–950 (1996)
7. Lin, L.-C., Chang, M.-C., Yang, H.C., Wei, M.-C.: Enhancing disaster prevention learning through human-centered design: Students' learning cognition and enjoyment in informal educational settings. Int. J. Disaster Risk Reduction **106**, 104451 (2024)
8. Chronis, A.: Coconstructing heritage at the gettysburg storyscape. Ann. Tour. Res. **32**(2), 386–406 (2005)
9. Park, H.Y.: Heritage tourism. Ann. Tour. Res. **37**(1), 116–135 (2010)
10. Ennes, M., Lee, I.: Distance learning in museums: a review of the literature. IRRODL. **22**(3), 162–187 (2021)
11. Broadus, R.N.: Toward a definition of "bibliometrics." Scientometrics **12**(5–6), 373–379 (1987)
12. He, Y., Yang, S., Chan, C.-Y., Chen, L., Wu, C.: Visualization analysis of intelligent vehicles research field based on mapping knowledge domain. IEEE Trans. Intell. Transport. Syst. **22**(9), 5721–5736 (2021)
13. Villatte, G., Marcheix, P.-S., Antoni, M., Devos, P., Descamps, S., Boisgard, S., Erivan, R.: Do bibliometric findings differ between Medline, Google Scholar and Web of Science? Bibliometry of publications after oral presentation to the 2013 and 2014 French Society of Arthroscopy (SFA) Congresses. Orthop. Traumatol. Surg. Res. **106**(8), 1469–1473 (2020)
14. Chen, C.: CiteSpace II: Detecting and visualizing emerging trends and transient patterns in scientific literature. J. Am. Soc. Inf. Sci. **57**(3), 359–377 (2006)
15. Azam, A., Ahmed, A., Wang, H., Wang, Y., Zhang, Z.: Knowledge structure and research progress in wind power generation (WPG) from 2005 to 2020 using CiteSpace based scientometric analysis. J. Clean. Prod. **295**, 126496 (2021)
16. Hollebeek, L.D., Sharma, T.G., Pandey, R., Sanyal, P., Clark, M.K.: Fifteen years of customer engagement research: a bibliometric and network analysis. JPBM. **31**(2), 293–309 (2022)
17. Benton, D.C., Benton, A.D., González-Jurado, M.A., Del Pulgar, M.G.: Comparing the scholarship contributions of five disciplines to the occupational licensure literature indexed in web of science^{TM}. J. Nurs. Regul. **10**(3), 4–13 (2019)
18. Sánchez, A.D., Cruz La, De., Del Río Rama, M., García, J.Á.: Bibliometric analysis of publications on wine tourism in the databases Scopus and WoS. Eur. Res. Manag. Bus. Econ. **23**(1), 8–15 (2017)
19. Vasconcelos, M.W., Gonçalves, S., Oliveira, E.C. de, Rubert, S., Ghisi, N. de C.: Textile effluent toxicity trend: a scientometric review. J. Cleaner Prod. 366, 132756 (2022)
20. Waltman, L., van Eck, N.J., Noyons, E.C.M.: A unified approach to mapping and clustering of bibliometric networks. J. Informet. **4**(4), 629–635 (2010)
21. Li, M., Wang, Y., Xue, H., Wu, L., Wang, Y., Wang, C., Gao, X., Li, Z., Zhang, X., Hasan, M., Alruqi, M., Bokhari, A., Han, N.: Scientometric analysis and scientific trends on microplastics research. Chemosphere **304**, 135337 (2022)
22. Antón, C., Camarero, C., Garrido, M.-J.: Exploring the experience value of museum visitors as a co-creation process. Curr. Issue Tour. **21**(12), 1406–1425 (2018)
23. Bideci, M., Albayrak, T.: An investigation of the domestic and foreign tourists' museum visit experiences. IJCTHR. **12**(3), 366–377 (2018)
24. Su, Y., Teng, W.: Contemplating museums' service failure: extracting the service quality dimensions of museums from negative on-line reviews. Tour. Manage. **69**, 214–222 (2018)
25. Brida, J.G., Disegna, M., Scuderi, R.: Visitors of two types of museums: a segmentation study. Expert Syst. Appl. **40**(6), 2224–2232 (2013)

26. Miles, W.F.S.: Auschwitz: museum interpretation and darker tourism. Ann. Tour. Res. **29**(4), 1175–1178 (2002)
27. Carter, P.L.: Where are the enslaved?: TripAdvisor and the narrative landscapes of southern plantation museums. J. Herit. Tour. **11**(3), 235–249 (2016)
28. Andermann, J.: Returning to the site of horror: on the reclaiming of clandestine concentration camps in Argentina. Theory Cult. Soc. **29**(1), 76–98 (2012)
29. Ashton, C.: Peking duck as a museum spectacle: staging local heritage for Olympic tourism. J. Tour. Cult. Chang. **10**(2), 150–166 (2012)
30. Baker, S., Cantillon, Z.: Safeguarding Australia's community heritage sector: a consideration of the institutional wellbeing of volunteer-managed galleries, libraries, archives, museums and historical societies. Aust. Hist. Stud. **51**(1), 70–87 (2020)
31. Kellett, P., Hede, A.-M.: Developing a sport museum: the case of tennis australia and the tennis heritage collection. Sport Manage. Rev. **11**(1), 92–117 (2008)
32. Mayfield, M.I.: Children's museums: purposes, practices and play? Early Child Develop. Care **175**(2), 179–192 (2005)
33. Waterton, E., Dittmer, J.: The museum as assemblage: bringing forth affect at the Australian War Memorial. Mus. Manag. Curatorship **29**(2), 122–139 (2014)
34. Phaswanamafuya, N., Haydam, N.: Tourists' expectations and perceptions of the Robben Island Museum—a world heritage site. Mus. Manag. Curatorship **20**(2), 149–169 (2005)
35. Quan, S., Wang, N.: Towards a structural model of the tourist experience: an illustration from food experiences in tourism. Tour. Manage. **25**(3), 297–305 (2004)
36. Shao, J., Ying, Q., Shu, S., Morrison, A.M., Booth, E.: Museum tourism 2.0: experiences and satisfaction with shopping at the National Gallery in London. Sustainability **11**(24), 7108 (2019)
37. Di Pietro, L., Guglielmetti Mugion, R., Mattia, G., Renzi, M.F.: Cultural heritage and consumer behaviour: a survey on Italian cultural visitors. J. Cult. Herit. Manag. Sustain. Development. **5**(1), 61–81 (2015)
38. Antón, C., Camarero, C., Garrido, M.-J.: A journey through the museum: visit factors that prevent or further visitor satiation. Ann. Tour. Res. **73**, 48–61 (2018)
39. Moreno Gil, S., Ritchie, J.R.B., Almeida-Santana, A.: Museum tourism in Canary Islands: assessing image perception of Directors and Visitors. Mus. Manag. Curatorship. **34**(5), 501–520 (2019)
40. Xu, M., Kim, S., Reijnders, S.: From food to feet: analysing *A Bite of China* as food-based destination image. Tour. Stud. **20**(2), 145–165 (2020)
41. Jansen-Verbeke, M., van Rekom, J.: Scanning museum visitors. Ann. Tour. Res. **23**(2), 364–375 (1996)
42. Cox, C., Wray, M.: Best practice marketing for regional tourism destinations. J. Travel Tour. Mark. **28**(5), 524–540 (2011)
43. Yang, C.-H., Lin, H.-L., Han, C.-C.: Analysis of international tourist arrivals in China: the role of World Heritage Sites. Tour. Manage. **31**(6), 827–837 (2010)
44. Leiper, N., Park, S.-Y.: Why the Gunarsa Art Museum fails as a tourism attraction. Tourism Culture Commun. **11**(2), 71–82 (2011)
45. Colombino, U., Nese, A.: Preference heterogeneity in relation to museum services. Tour. Econ. **15**(2), 381–395 (2009)
46. Çela, A., Lankford, S., Knowles-Lankford, J.: Visitor spending and economic impacts of heritage tourism: a case study of the Silos and Smokestacks National Heritage Area. J. Herit. Tour. **4**(3), 245–256 (2009)
47. Tolia-Kelly, D.P.: *Feeling* and *Being* at the (Postcolonial) museum: presencing the affective politics of 'Race' and Culture. Sociology **50**(5), 896–912 (2016)
48. Gieling, J., Ong, C.-E.: Warfare tourism experiences and national identity: the case of airborne museum 'Hartenstein' in Oosterbeek, the Netherlands. Tour. Manage. **57**, 45–55 (2016)
49. Serravalle, F., Ferraris, A., Vrontis, D., Thrassou, A., Christofi, M.: Augmented reality in the tourism industry: a multi-stakeholder analysis of museums. Tourism Manage. Perspect. **32**, 100549 (2019)

50. Tan, F., Gong, X., Tsang, M.C.: The educational effects of children's museums on cognitive development: empirical evidence based on two samples from Beijing. Int. J. Educ. Res. **106**, 101729 (2021)
51. Iannone, F., Izzo, F.: Salvatore Ferragamo: an Italian heritage brand and its museum. Place Brand. Public Dipl. **13**(2), 163–175 (2017)
52. Farmaki, A., Antoniou, K.: Politicising dark tourism sites: evidence from Cyprus. WHATT. **9**(2), 175–186 (2017)
53. Zhang, C.X., Xiao, H., Morgan, N., Ly, T.P.: Politics of memories: identity construction in museums. Ann. Tour. Res. **73**, 116–130 (2018)
54. Visser, M., van Eck, N.J., Waltman, L.: Large-scale comparison of bibliographic data sources: Scopus, Web of Science, Dimensions, Crossref, and Microsoft Academic. Quant. Sci. Stud. **2**(1), 20–41 (2021)

Influence Factors of Pro-Poor Tourism on the Perception of Ethnic Minority Women

Mengtian Cao and Xialei Duan

Abstract Poverty feminization has become a major obstacle to the sustainable livelihood development of ethnic minority farmers in China. This study examines the Yao Autonomous County of Liannan in Guangdong Province as a case study. We employ a hierarchical linear model to analyze key factors influencing local women's pro-poor tourism perception at individual, family, and rural community levels. The hierarchical linear model effectively addresses data correlation, imbalance, and missing panel data, resulting in more stable and accurate estimations. Our results reveal that key factors at individual and family levels include women's monthly income, number of family members participating in tourism, and maximum years of education among working family members, all showing positive effects. At the rural community level, key factors affecting pro-poor tourism perception include village poverty, pro-poor tourism project indicators, and whether the village is exclusively Yao. The pro-poor tourism project indicator shows a positive effect, while other indicators demonstrate negative effects. Based on these findings, we present policy recommendations.

Keywords Women · Ethnic minority · Pro-poor tourism · Perception · Liannan county

1 Introduction

Women's poverty exists in every country in the world. Eliminating the poverty of women and breaking the intergenerational transmission of poverty has become one of the problems that should be solved urgently in the world [1, 2]. Like age, ethnic factors

M. Cao
Institute of Tourism Management and Planning and Design, Institute of Lingnan Tourism Research, Guangdong University of Finance & Economics, Guangdong University of Finance & Economics, Guangzhou, Guangdong, People's Republic of China

X. Duan (✉)
International School of Tourism and Management, City University of Macau, Avenida Padre Tomás Pereira Taipa, Macau, People's Republic of China
e-mail: sallyduan@cityu.edu.mo

Y. Luo et al. (eds.), *Tourist Experience and Perception in the Pandemic and Post-Pandemic Era*, https://doi.org/10.1007/978-981-96-1820-0_11

and geographical location, the possibility of poverty is not randomly distributed among the population, and gender is an important factor affecting poverty, so the research on anti-poverty cannot be separated from the perspective of gender [3]. According to the statistics of the United Nations, 70% of the world's poor were women at the end of 2015, and compared with men, they suffered higher risks of poverty and were more vulnerable [4, 5]. "Poverty feminization" was first put forward in the United States in the late 1970s. Pearce [6] has believed that there are obvious gender differences in poverty, arguing that "poverty feminization" mainly refers to the increasing proportion of female population in poverty-stricken population and the increasing proportion of female-headed households in all poverty-stricken households. For example, the proportion of female-headed households in the United States increased rapidly from 28% in 1960 to 60% in 1987 [7]. In addition, in many countries, it has become normal for both husband and wife to participate in the labor market. Therefore, female-headed households with only one earner are vulnerable [8, 9] According to the social survey data of 14 poverty-stricken villages from Gansu Province in China, it is seen that the incidence of unmarried women's poverty is higher, and compared with unmarried women, the probability of family poverty of married women is reduced by 48.5% [10].

From a global perspective, poverty has a female face, which mainly comes from unemployed women, low-income women, migrant women, landless women farmers, female-headed single-parent households, elderly women, HIV-infected women and a new type of female poverty caused by climate change disasters. These women have multiple roles such as child rearing, family production and community participation, and they are special in terms of gender, occupation, community participation, economic development and family life [11, 12]. This is due to the constraints of women's physiology, society and other subjective and objective factors, and for example, the labor participation rate of women has been greatly improved, and the traditional mode of "bread winning men and homemaking women" has been gradually broken. However, with the exception of a few countries, the labor participation rate of women is still obviously lower than that of men, and women's economic income and status have not been significantly improved [13]. According to the survey, it is seen that in many countries, the employment rate of women without childbirth history is equal to that of men, and once women become mothers, their employment rates and wage levels will drop significantly. Due to these factors, women have been excluded from high-paying jobs in the primary labor market. Low-paying and unstable jobs often lead to low levels of pension security for women in their old age, and women's poverty will continue into the old age [14]. The unequal distribution of resources within the family increases the possibility of women's poverty. Traditionally, the family is regarded as a homogeneous system, so women's poverty is gender-blind and difficult to detect. For instance, in the mid-1990s, about a quarter of women in France, the United Kingdom, the United States and Germany had less than one-tenth of the family's entire resources in their own names [15, 16]. With the change of people's concept of traditional marriage and family, the situation of divorce, separation and unmarried childbirth has increased greatly, resulting in many single-parent female-headed households composed of women and their minor children. According to the research on gender differences in Australia, Canada, France, Germany, the Netherlands, Sweden, the United Kingdom and the United States, it is found that single mothers are at a higher risk of poverty, while these families where women raise their minor children independently are characterized by intergenerational transmission of poverty [17, 18]. This is because divorce and single parents are culturally excluded in society, and women are mostly at a disadvantaged in the division of property, so that they are often treated unfairly and marginalized [19, 20]. As a nation with the world's largest poverty reduction population, China has contributed more than 70% to the global poverty reduction. In 2018, China's rural poverty-stricken population has decreased from 98.99 million at the end of 2012 to 16.6 million, with a cumulative reduction of 82.39 million, and the incidence of poverty has decreased from 10.2% in 2012 to 1.7%, with a cumulative decline of 8.5% [21]. According to the existing registration data of poverty-stricken population in China, it is seen that with regard to income, the incidence of women's poverty is 0.4% higher than that of men's poverty. In terms of education level, the proportion of illiteracy in female laborers is 10.2% higher than that of illiteracy in male laborers. Especially in ethnic minority areas, women's poverty has a profound impact on the sustainable development of the local rural areas and the construction of the new countryside [22]. At the same time, women's poverty is characterized by intergenerational transmission, diversity, vulnerability and concealment. Regarding personal subjective factors, women's physiological basis, personal capital and ideological limitations are the main factors leading to poverty. Since women are more active in the unpaid economy, many women's contributions to the family (childbirth, care, housework, etc.) are regarded as natural and unpaid "labor of love" [23, 24]. Concerning social structure factors, women's limited selectivity in the face of labor market nature, social distribution system, social security system and other structural factors is the structural factor that leads to poverty. As far as environmental factors are concerned, the main causes of women's poverty include natural environment, institutional environment and social environment.

Since the 1980s, pro-poor tourism, a special form of "the tourism demand", has sprung up, helping to reduce poverty, hunger, gender inequality and environmental degradation, which threaten the world [25]. In 2002, United Nations World Tourism Organization (UNWTO) began to organize and implement "Sustainable Tourism-Eliminating Poverty" (ST-EP). As the content of tourism poverty alleviation continues to expand, the research involves the relationship among tourism managers, investors, tourists and other relevant interest groups. Therefore, scholars prefer to use ST-EP instead of pro-poor tourism. The core of ST-EP is to take sustainable tourism as a means of poverty reduction and use the high-quality tourism resources of underdeveloped countries and developing countries to drive the local poor people to participate in tourism management services to achieve poverty alleviation. In particular, women participate in economic development more effectively, and the number of female employees increases. Tourism employment has improved women's economic independence, decision-making ability and social freedom, which brings about the positive improvement of gender relations in family and community environment. For example, in the tourism industry, even 90% of the employees engaging in the catering services and hotel accommodation services are women, and tourism provides women with a broad space for development, playing an important role in reducing women's poverty [26, 27]. However, in China, the continuous transfer of rural male laborers to the cities has resulted in numerous left-behind female laborers in rural areas and ethnic minority areas, marked problems of "surplus labor feminization", women's low employment levels and high-income poverty rates. Rural women and ethnic minority women constitute the bottom of the poor [28]. At the same time, the characteristics of tourism feminization are prominent. As the main force of tourism employees and the important part of the poor people, women are most likely to participate in the local rural pro-poor tourism and have the highest pro-poor tourism perception. With the continuous development of rural pro-poor tourism, it is necessary to study the influencing factors of the local women's pro-poor tourism perception and pertinently carry out pro-poor tourism activities to allow more women in ethnic minority areas to benefit from them, so as to achieve poverty reduction.

Therefore, Gender-sensitive researchers have begun to pay more attention to women's tourism poverty alleviation. The research on gender pro-poor tourism focuses on the relationship between pro-poor tourism and women, and special attention should be paid to families with female household heads in developing countries in terms of enhancing women's personal values, because they are considered to be in an unfavorable position in terms of obtaining land, laborers, credit and insurance markets. Women with different demographic characteristics (such as age and educational level) have different levels of pro-poor tourism perception and different attitudes towards pro-poor tourism perception. Through the research, the suggestions for the influence and management of the local environment are put forward, to guide the sustainable development of the local pro-poor tourism [29, 30]. Sainsbury and Morissens [31] believe the poverty rate can also be effectively reduced by increasing their participation in tourism and improving their working conditions. To some extent, it can result in the change of rural households' family gender relations and promote the equality of family gender relations. The research from Kwaramba has pointed out that tourism poverty reduction is an important part of the South African government's program to promote local economic development. The purpose of promoting tourism home stay is to help disadvantaged women become self-reliance entrepreneurs and use the emotional intelligence method to assess women's rights and confidence levels. The research has shown that disadvantaged groups and low emotional confidence levels can strive to grasp the opportunities of economic development in order to achieve regional economic development and thereby reduce poverty [32]. Von der Weppen and Cochrane [33] research points out that women playing the role of "social entrepreneurs" use the training methods to create employment opportunities, which play a certain role in promoting pro-poor tourism and environmental protection. Studying the challenges faced by women in the traditional tourism industry in the Dominican Republic, Duffy has surveyed the residents' gender awareness and the relationship between tourism and women's employment in 12 coastal communities and has pointed out that tourism provides women with employment opportunities so as to achieve economic and social independence [34]. Tourism is very suitable for rural women, because they cannot only consider housework and agricultural production, but also obtain employment opportunities and get extra incomes. Their attitudes and perceptions of tourism are constantly evolving as well [35].

The research method has gradually shifted from the early qualitative analysis of the economic performance and multiplier effect of pro-poor tourism to today's extensive combination method of the qualitative analysis and the quantitative analysis. In fact, the "gender-sensitive" poverty standard created by Jaggar is a more realistic and gender-focused research method, which means communicating with the poor equally, paying attention to the situation and expression of women all the time in the whole process of research and analyzing and comparing them with the equivalent aspect of men in the same situation. This can not only comprehensively and deeply investigate women's poverty but also reveal the systematic gender injustices they have encountered, to reveal the important connotation and basic characteristics of "poverty feminization" [36]. Among them, the qualitative analysis is dominated by social surveys, and most scholars have adopted questionnaire surveys, Delphi method, inductive method and interviews to understand the residents' judgments of some measurement indicators in tourist sites and try to subdivide and quantify their perceptions of tourism impact [37]. Constructing the measurement methods of pro-poor tourism perception based on the social gender mainstreaming, formulating the preferential policies that are beneficial to eliminating women's poverty and establishing the gender-based tourism employment statistics indicators is conducive to accelerating rural women's poverty reduction. Meanwhile, it is also a mainstream research method to select the case site and establish the indicator system for quantitative analysis. For example, EQ analysis, factor analysis, principal component analysis, tourism multiplier effect, general equilibrium model, geographical weighted regression model, incremental analysis, entropy method and TOPSIS model are used to explore and measure the local residents' pro-poor tourism perception [38]. Guo [39] has used the industrial chain tracking analysis method to study the impact of tourism industry scale and tourism development on poverty-stricken population, the combined effect of tourism and related industries and the competitiveness of tourism. From the perspective of industrial chain, the commonly used research methods include input-output analysis, tourism satellite account analysis, regression analysis, social accounting matrix method, computable general equilibrium model, etc., which can be used as the basic methods for the future research. Many scholars have used the data envelopment method to study the efficiency of tourism poverty alleviation and the time evolution of efficiency change. Through ArcGIS and other drawings, they have illustrated the spatial changes of tourism poverty alleviation efficiency in the case area, putting forward the corresponding suggestions for tourism poverty alleviation [40].

To sum up, in the current studies of women's tourism poverty alleviation, most of them have sought to explore various factors that lead to poverty, including natural environmental factors, socio-economic limitations, institutional cultural constraints, etc., but have failed to put rural women's poverty into a comprehensive and systematic discussion of livelihood vulnerability. In terms of anti-poverty strategies, they have mainly focused on improving rural women's cultural level and quality, enhancing their right consciousness and feasible capacity and changing their weak position through the efforts of the government and non-governmental organizations, while ignoring the initiative and subjectivity of rural poor women as the action subject within the region. From the multi-layered nature of the study, various studies have been conducted generally at the microscopic individual and family levels or the macroscopic community level, which is relatively single, separates the nested relationship of female individuals and families in rural communities and cannot systematically and comprehensively analyze the individual, family and community as a whole. Thus, firstly, based on the previous studies, this study takes rural women in Liannan County as the research object. Meanwhile, through literature review, questionnaire surveys and field interviews, it understands women's overall perception of pro-poor tourism from the microscopic perspective of their own perceptions. Furthermore, through establishing the complete hierarchical linear model, it analyzes the key influencing factors of pro-poor tourism perception, putting forward some relevant suggestions for the precise rural pro-poor tourism in accordance with the research results.

2 Study Area and Data Sources

2.1 Study Area

With an area of 1,306 km2, Liannan County (24°17′–24°56′N, 112°2′–112°29′E) has jurisdiction over seven towns. It is characterized by more mountains, less land, favorable ecological environment and significant karst landscape, belonging to one of the impoverished ethnic minority counties in Guangdong Province of China. The county had a total population of 173,727, including 90,507 males and 83,222 females, which accounted for 52.10% and 47.90% of the total population, respectively. The Yao population was 95,079, accounting for 54.73% of the total population. Liannan County is an autonomous county with the largest population of ethnic minorities in Guangdong Province and the only agglomeration area of rows of Yao villages in the world. At the same time, there were 5,930 people from 2,171 registered poor households (including 4,599 people from 1,228 poor households with working capacities and 1,331 people from 943 poor households without working capacities), which consisted of 2,329 people from 636 general poor households, 3,033 people from 1,000 low-income households and 568 people from 535 five-guaranteed households. At the end of 2018, in the county, the annual per capita disposable income of residents was18,066.8 yuan, an increase of 8.2 percent, with the average per capita income of 12,262 yuan per poor households with working capacities [21].

With a total of 133 landscapes, the county is endowed with abundant and diverse tourism resources, and the cultural tourism resources of the Yao nationality are the most prominent. There are two five-star resources (namely Harvest Celebration Festival and Youling Rows of Ancient Villages) and eight four-star tourism resources (namely Millennium Yao Village, Embroidery of the Yao Nationality,

Long-Drum Dance of the Yao Nationality, Museum of the Yao Nationality, Bonfire Party, Numerous Mountains' Worship, Shiyangkeng Mineral Warm Spring and Wild Giant Salamander Reserve) in the county. At present, the tourist attractions that have been opened include Millennium Yao Village (4A), Museum of the Yao Nationality (4A), Youling Folk Yao Village and Numerous Mountains' Worship Tour Station.

Liannan County has 3 state-level ancient villages and 7 state-level distinct ethnic minority villages, and Youling Yao people's residential group has been successfully declared as the provincial cultural relic protection unit. "Panwang Festival", "Harvest Celebration Festival", "Ploughing Festival", "Singing Festival", "Paddy Field Fish Festival" and other Yao festival activities as well as "eight rows of Yao's marriage customs", the tie-dyeing of the Yao nationality and the making techniques of Pai Yao cowhide pastry are included in the provincial intangible cultural heritage list. There are 9 intangible cultural heritage projects above the provincial level, and the number of inheritors of intangible cultural heritage above the provincial level has increased to 11. Youling Village has been rated as the base for inheriting the long-drum dance and the long-drum production techniques of Yao nationality of the intangible cultural heritage in Guangdong Province, and Guangdong Museum of Yao Nationality has been named as the base for inheriting Liannan Yao people's clothing embroidery of the intangible cultural heritage in Guangdong Province. At the same time, embroidery of Yao nationality, silver ornaments of Yao nationality and other distinctive cultural industry products are very abundant.

Therefore, regarding the development of tourism as the main means of the precise poverty alleviation to promote the local economic development has achieved remarkable results. In 2018, the total number of tourists was 3.09 million, with a year-on-year increase of 10.80%, and the total tourism revenue was 1.19 billion Yuan, with a year-on-year growth of 10.9%. It has helped nearly 16,000 people find jobs and start businesses. The number of domestic and foreign tourists visiting the county through the county's travel agencies was 45,620, with a year-on-year increase of 19%, and the total number of domestic and foreign tourists living in the county's hotels, guesthouses and social hotels was 900,000. The tourism pattern of the whole region has gradually formed, and the output value accounts for a third of the county's GDP. It has become the pillar industry of Liannan County.

2.2 Data Sources

The data used in this paper come from the survey data of "The Research on pro-poor tourism perception of Rural Households in Northern Guangdong Province" conducted by Lingnan Tourism Research Academy, Guangdong University of Finance & Economics in July 2017 to July 2020. The specific items of the survey are shown in Table 1. The sampling method of the samples adopts the stratified sampling and the random sampling, and the stratified sampling indicator system for the sample county is constructed, including per capita GDP of towns, net income of rural residents, environmental conditions of tourism resources, status of tourist

Table 1 The investigation item at all levels in Liannan County

Individual level investigation item	Family level investigation item	Rural community level investigation item
Women's age	Family structure characteristics	Rural pro-poor tourism projects
Women's education	Household labor conditions	The development of rural tourism
Women's occupation and monthly income	Household capital position	Construction of rural transportation lines
Women's relationship with tourism	Household income and expenditure	Rural infrastructure construction
Women's pro-poor tourism perception	Household livelihood risk	The state of rural ethnic culture

attractions, traffic accessibility, accessibility of Yao residents and other indicators. Through the principal component analysis and the hierarchical clustering method, 71 villages from 7 towns in the county are divided into 8 groups, and one sample village is randomly selected from each group, so 8 sample villages are obtained. The relatively strong homogeneity of rural households living in the same rural community is taken into account, the questionnaire surveys are conducted among 45 women respectively from 45 rural households, which are randomly selected from each sample village. A total of 360 samples are obtained. Through analysis and arrangement, after the removal of the invalid questionnaires, 329 valid questionnaires are obtained, and the valid rate reaches 91.39%. The distribution of the sample county and the sample villages is shown in Fig. 1. The project survey method of the sample rural communities adopts the participatory evaluation survey method. The informal interviews of the main cadres in the county, towns and villages are conducted, so as to obtain the general and special characteristics of the current status of the socio-economic development of the ethnic autonomous county and understand the tourism resources in rural communities, the national culture characteristics, the current status and policies of rural pro-poor tourism, etc.

3 Design of Pro-Poor Tourism Perception Indicators and Model Methods

3.1 Design of Indicators and Analysis of Research Hypothesis

Overall, the investigation of women's pro-poor tourism perception in Liannan County mainly includes three indicators, which are the advantages and disadvantages of the effect of developing rural pro-poor tourism in rural communities, being satisfied with

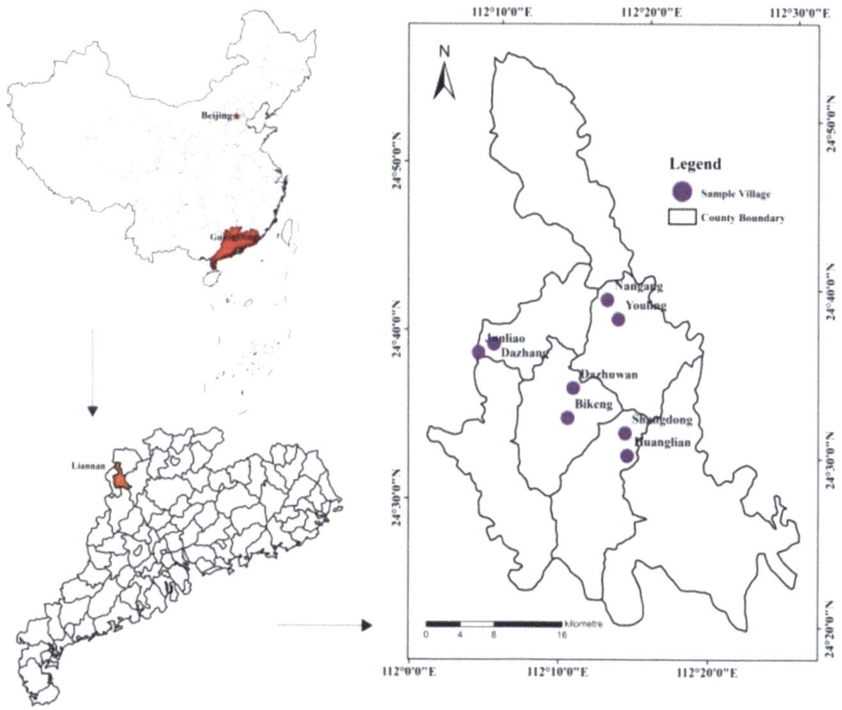

Fig. 1 The distribution of the sample county and the sample villages in Liannan county

the current status of developing rural tourism in rural communities and being willing to support the development of rural tourism in the village (1 = strongly disagreeable, 2 = disagreeable, 3 = general, 4 = agreeable, 5 = strongly agreeable). The mean value of the three indicators is taken as the dependent variable. The cross-section of 329 sample women's pro-poor tourism perception data is presented in Fig. 2. By calculation, the average eigenvalue of women's overall pro-poor tourism perception in Liannan County is 3.80, and the maximum value and the minimum value are 5.00 and 1.00, respectively.

3.2 *The Personal Dimension of Women*

Based on the actual situation of the county, the environmental influencing factors at women's individual and family levels and the rural community level are considered comprehensively in this study. The indicators of the model are designed as follows.

Women's age (WA). With the acceleration of the aging process of the population, women live longer than men in terms of physiological characteristics, and elderly

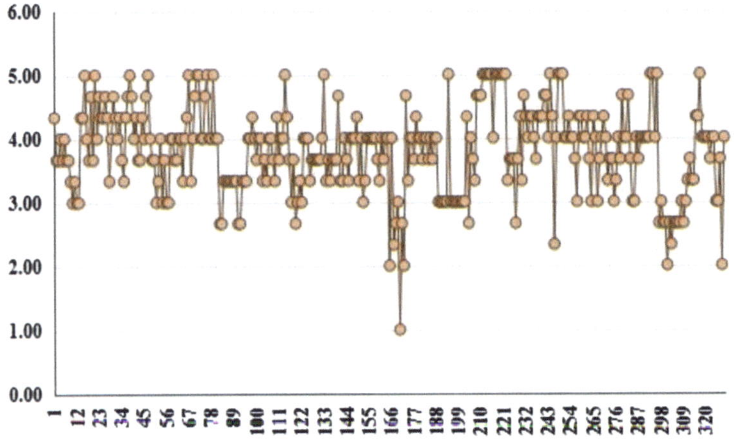

Fig. 2 The cross-section of 329 sample women's Pro-poor tourism perception data

women account for a large proportion. Elderly women have to suffer the psychological pressure and loneliness caused by widowhood, so they are more willing to participate in the tourism industry and expand the circle of friends. This study assumes that if women are older, their pro-poor tourism perception will be higher, and they are more likely to participate in tourism.

Women's education (WE). In modern society, the education level of women is increasing year by year, so women have more and more opportunities to engage in formal occupations, and more and more women devote themselves to the workplace. This study assumes that if their educational levels are higher, their pro-poor tourism perception will be higher, and they are more likely to participate in tourism.

Whether women are engaged in agricultural production (WP). Whether the female respondents participate in household agricultural production work, including planting crops, breeding livestock, etc. Generally speaking, rural labor women, especially middle-aged women, need to stay at home to take care of the elderly and children, and they have too much housework to do, so they have no time to deal with non-agricultural employment. This study assumes that women engaged in family agricultural production have a lower perception of tourism poverty alleviation and are less likely to participate in tourism.

Whether women master skills (Whether the female respondents master some skills, including medical skills, driving skills, cooking skills, etc.). This study assumes that women have a certain skill, and their willingness to participate in the public affairs of tourism poverty alleviation is not strong, resulting in a lower perception of tourism poverty alleviation.

Whether women are engaged in tourism (WT). Whether the female respondents participate in tourism, including hotels, catering, tour guides, scenic shows, etc. They have a better understanding of tourism and are more involved in tourism, so this study assumes that women engaged in tourism have a higher perception of tourism poverty alleviation.

Women's monthly incomes (WM). Women's monthly income is also an important influencing factor. Women have lower monthly incomes are more eager to participate in tourism to increase income, and their pro-poor tourism perception is higher.

3.3 *Factors in Women's Families*

From the perspective of sustainable livelihoods, the poverty status of local women has the general characteristics of poverty-stricken population, but still has its own particularities, which exist in the family, community and other aspects. The lack of livelihood capital and vulnerability are the main obstacles for them to get rid of poverty. Concerning the livelihood asset pentagon of the framework for rural households' sustainable livelihoods based on DFID, rural households' natural, social, physical, financial and human capitals are selected to design the relevant indicators.

Natural capital is most closely related to vulnerability. The geographical location, climate environment and natural resources of a region largely determine the amount of natural capital that people have and the risks that they face. Poor natural environment conditions determine that poverty-stricken women are faced with more risks and uncertainties. Agricultural production is the basic source of livelihoods for rural households in Liannan County. Family's cultivated land area (FC) is selected in this paper. This study assumes that if the family's cultivated land area is larger, left-behind women at home will need to spend more time on farm work, so they will be less likely to participate in tourism, and their pro-poor tourism perception will be lower.

Physical capital is represented by means of production, and its lack directly affects the improvement of productivity and production efficiency. The present value of the family's houses (HV) and the present value of the family's durable consumer goods (the current sales value of the family's durable consumer goods, including automobiles, refrigerators, televisions, large furniture, etc.)(FD) are selected in this paper. This study assumes that the higher present value of the family's houses and the higher present value of the family's durable consumer goods will result in the stronger liquidity, the smaller possibility of returning to poverty and the lower pro-poor tourism perception.

Social capital is the social networks that people have and the ability to obtain resources through those networks. People can obtain various tangible or intangible, material or non-material help through social capital, and then implement livelihood strategies to provide opportunities for the development of individuals or families. The scarcity of social capital can weaken the communication and cooperation between female individuals and the outside world. The number of family members participating in tourism (FM) and whether women have relatives or friends who participate in tourism (FR) are selected in this paper to indirectly reflect the degree of the social connection between women's families and the outside world. For example, family members, relatives and friends directly engaging in tourism can provide women with information on pro-poor tourism and tourism employment to help women improve pro-poor tourism perception, so if more family members participate in tourism, and women have relatives or friends who directly participate in tourism, women will have higher pro-poor tourism perception.

Financial capital means the accumulating and circulating funds that individuals need in order to achieve livelihood goals in the process of consumption and production. It mainly refers to the family's disposable capital reserves and various funds obtained from formal and informal channels. The shortage of financial capital indicates the scarcity of capital stock and flow.

The amount of the family's yearly income (FY) directly determines the poverty level of the family, and the family's net income, including agricultural net income and non-agricultural net income, is mainly used to calculate the family's yearly income in this paper. Agricultural net income comes from the production of agricultural products and is converted in accordance with the local market prices, such as subtracting the related costs of the purchase of seeds, seedlings, baby livestock, livestock feed, fertilizers, pesticides and so on from the production of rice, corn, tea, fruit tree planting, poultry and livestock breeding and other agricultural products. Non-agricultural family income is derived from the net income obtained from the secondary industry and the tertiary industry, such as the remaining profits after deducting the costs from the incomes of canteens, tourism catering, homestays and agriculture entertainment tourism projects. This study assumes that the higher family's yearly income will lead to the lower vulnerability and the lower pro-poor tourism perception. In general, farmers' families take measures of reducing consumption and using savings and loans to cope with diseases, disaster risks and other shocks. As indicated from the survey, rural households in Liannan County have almost no savings or bank loans, and in the event of economic difficulties, they would choose to borrow money from relatives and friends (FB, the amount of the family's borrowings from relatives and friends). This study assumes that more borrowings in the past three years will lead to the higher vulnerability, the higher pro-poor tourism perception and the greater desire to participate in tourism.

As a form of capital relative to physical capital, human capital manifests mainly as personal knowledge, skills and abilities. In the livelihood capital system, the quality of human capital owned by individuals determines whether they can use other capital to achieve the corresponding livelihood goals. The lack of human capital seriously hinders their livelihood development. The scale of the family's laborers (the number of people with working capacities in the family) (FS) and the maximum years of education of the laborers in the family (the highest academic qualification of the person with working capacities in the family) (FE) are selected in this paper. This study assumes that the larger scale and the higher educational level of the family's laborers will result in the higher pro-poor tourism perception.

3.4 Factors at the Level of Rural Community

At the rural community level, the level of women's pro-poor tourism perception is influenced not only by their individual and family factors, but also by the rural community factors [41]. In the process of tourism poverty alleviation, there is a need to pay more attention to local communities. Concerning the indicators of the sample rural communities, whether the village is the poor village (PV), whether there is a rural pro-poor tourism project (PP) and whether the village is the pure Yao village (VY) are selected in this paper to examine the situation of the macroscopic policies in China. This study assumes that the women in rural communities with pro-poor tourism projects and preferential policies for ethnic minorities have higher pro-poor tourism perception. The rural communities that are closer to the county (DC) are economically radiated and promoted by the center areas of the county, and the villages have the more perfect infrastructure and the basis for tourism development, so pro-poor tourism perception is higher; the distance from tourist attractions (DA) (the distance between the rural community and the nearest tourist attraction) affects women's employment in the scenic areas. This study assumes that if the distance from tourist attractions is longer, the commuting and freight costs for women's participation in employment as well as sales of handicrafts in scenic areas will be higher, and the risks of participation in tourism will be higher, but pro-poor tourism perception will be lower. The symbols and definitions of the indicators at various levels are shown in Table 2.

3.5 Model Methods Principle

According to the design of the above model indicators, this study constructs the complete two-level linear model at women's individual and family levels and the rural community level in Liannan County. The model principle is OLS regression estimation of the indicators at women's individual and family levels based on Level-1 as well as WLS contraction estimation of the indicators at the rural community level based on Level-2, which make the estimation results more stable and accurate. The specific formulas of the hierarchical linear model are as follows:

$$\text{Level 1}: PPTP_{ij} = \beta_{0j} + \beta_{ij}X_{nij} + \chi_j \tag{1}$$

$$\text{Level 2}: \beta_{0j} = \gamma_{00} + \gamma_{01}W_{nj} + \mu_{0j} \tag{2}$$

$$\beta_{1j} = \gamma_{10} + \gamma_{11}W_{nj} + \mu_{1j} \tag{3}$$

$$V_{ar}(\mu_{0jk}) = \tau\beta_{00} \tag{4}$$

Table 2 The symbols and definitions of the indicators at various levels

Indicators name	Symbol	Definition
Level-1		
Women's age	WA	Actual age of the female respondents. (year)
Women's Education	WE	Women's years of education. (year)
Whether women are engaged in agricultural production	WP	
Whether women are engaged in tourism	WT	Whether the female respondents participate in household agricultural production work, including planting crops, breeding livestock, etc. (1 = Yes;0 = No;)Whether the female respondents participate in tourism, including hotels, catering, tour guides, scenic shows, etc. (1 = Yes;0 = No;)
Women's monthly income	WM	The monthly income of female respondents. (104 yuan)
Whether women master skills	WS	Whether the female respondents master some skills, including medical skills, driving skills, cooking skills, etc. (1 = Yes; 0 = No;)
Family's cultivated land area	FC	The amount of cultivated land owned by a family. (mu)
Family members participating in tourism	FM	The number of family members participating in tourism
Family relatives and friends participating in tourism	FR	Whether the relatives and friends participate in tourism, including hotels, catering, tour guides, scenic shows, etc. (1 = Yes; 0 = No;)
The present value of the family's houses	FH	1 = 0–5; 2 = 5–10; 3 =≧ 10(104yuan)
The present value of the family's durable consumer goods	FD	The current sales value of the family's durable consumer goods, including automobiles, refrigerators, televisions, large furniture, etc.
Family's borrowings	FB	The amount of the family's borrowings from relatives and friends (104yuan)
Family's yearly income	FY	The amount of family's agricultural net income and non-agricultural net income
The scale of the family's laborers	FS	The number of people with working capacities in the family
Family's laborers' education	FE	The highest academic qualification of the person with working capacities in the family. (year)
Level-2		
Poor village	PV	Whether the village is the poor village. (1 = Yes; 0 = No)

(continued)

Table 2 (continued)

Indicators name	Symbol	Definition
Pro-poor tourism project	PP	Whether there is a rural pro-poor tourism project. (1 = Yes; 0 = No)
Pure Yao village	YV	Whether the village is the pure Yao village. (1 = Yes; 0 = No;)
The distance from county center	DC	The distance between rural community and county center. (km)
The	distance from tourist attractions DA	The distance between the rural community and the nearest tourist attraction. (km)

(*Note* [a]1 USD= 6.29 yuan (During the study period); [b]1 mu ≈ 667 m^2 or 0.667 ha.)

$$V_{ar}\left(\mu_{1jk}\right) = \tau\beta_{11} \quad (5)$$

$$COV = \left(\mu_{0jk}, \mu_{1jk}\right) = \tau\beta_{01} \quad (6)$$

where, i indicates the rural household; j stands for the rural community, X refers to the indicator at women's individual and family levels; pro-poor tourism perception$_{ij}$ represents the eigenvalue of women's pro-poor tourism perception; β_{0j} and β_{1j} denote the intercept and the slope of the first level related to unit j of the second level, respectively. γ_{00} andγ_{10} refer to the mean values of β_{0j} and β_{1j}, respectively, and are the fixed components of β_{0j} and β_{1j}, indicating that they are constant among rural community units; γ_{01} andγ_{11} stand for the regression slope. W denotes the prediction indicator at the rural community level; μ_{0j} and μ_{1j} represent the random components of β_{0j} and β_{1j}, indicating the differences among rural community units.

4 Analysis of the Model Results

4.1 Analysis on the Relevant Description of Female Respondents

4.1.1 Personal Characteristics of Female Interviewees

As can be seen from Table 3, the female respondents are mainly the young and middle-aged people, with the average age of 42.39, and women aged between 20 and 45 and women aged over 46 account for 67.48% and 32.52% of the total samples, respectively. The overall educational level of the female respondents is low, and the average years of education are 5.78, while women with primary school degrees, junior high school degrees and senior high school degrees and above account for 67.17%, 26.44% and 6.39% of the total samples, respectively. This is mainly because Liannan

Table 3 Level-1 indicators descriptive statistics

Indicators	Min	Max	Mean	SD	Indicators	Min	Max	Mean	SD
WA	20.00	65.00	42.39	9.11	FD	0.00	26.56	1.37	2.78
WE	0.00	22.00	5.78	3.60	FB	0.00	15.00	1.54	2.59
WP	0.00	1.00	0.71	0.46	FY	0.04	26.40	5.24	4.32
WT	0.00	1.00	0.57	0.50	FS	1.00	7.00	3.56	1.44
WM	0.00	0.80	0.071	0.14	FE	0.00	22.00	9.74	3.63
WS	0.00	1.00	0.35	0.48	PV	0.00	1.00	0.75	0.46
FC	0.00	40.00	2.87	4.07	YV	0.00	1.00	0.63	0.52
FT	0.00	5.00	0.22	0.57	PP	0.00	1.00	0.50	0.54
FR	0.00	1.00	0.23	0.42	DC	5.80	57.10	25.99	18.47
FH	0.00	100.00	14.24	14.38	DA	3.70	41.30	18.93	14.36

County is located in the remote mountainous area, and female laborers with higher educational levels mostly go out to work, whereas women with lower educational levels are mostly engaged in agricultural production and take care of their children at home. The average monthly income of women is 709.09 Yuan, and specifically, women with the monthly incomes of less than 200 Yuan and more than 1,000 Yuan account for 69.61% and 20.37% of the total samples, respectively, indicating that the local women's incomes are low, and women are still a vulnerable group among the poor people. 56.84% of the female respondents are engaged in tourism, and women's participation in the local tourism is at a moderate level. 34.95% of the women master some skills, suggesting that the local women's technical abilities are relatively low.

4.2 *Family Characteristics of the Women Interviewed*

Natural capital is scarce, and profits are relatively low. According to the survey, cultivated land is the main natural capital of farmers' families, and the average cultivated land area of the female respondents' families in Liannan County is 2.87 mu, the per capita cultivated land area is only 0.61 mu. Liannan County has poor land, rugged terrain and frequent floods and droughts, so it is impossible to carry out patch planting development and mechanical operation. Therefore, people can only rely on natural conditions to live, and both per unit yield and total yield of crops are extremely unstable. In the backward agricultural production process, women who play the roles of housewife, mother and wife together pay more labor than men, but earn less. Due to blocking traffic and backward information, rural women can only engage in simple agricultural production at home, which mainly focuses on the traditional planting and breeding industries.

The quality of life is poor, and the physical capital is scarce. The average present value of the female family's houses is 142,400 Yuan, and the average present value of the female family's durable consumer goods is 13,700 Yuan. The current value of per capita housing is 32,100 Yuan, per capita housing area is 24.90 square meters, per capita present value of durable consumer goods is only 2,953 yuan. In terms of housing, concrete houses and brick and tile houses account for 71.43% and 28.57%, respectively. Only a few families live on both sides of the village road, and most of the families live in dilapidated houses. The number of vehicles and agricultural production machinery is small, and there are significant differences in using them between men and women. Transport vehicles, motorcycles, communication tools and farming machinery are basically controlled by men, and women can only rely on men in the family, so they are in a relatively weak position. In the case where the family's physical capital is relatively scarce, the reality of gender inequality hidden in the family's shared resources intensifies the depth of women's poverty. Social networks are vulnerable, and social capital is scarce. 77.2% of the female respondents have no family members directly participating in tourism, and 44.16% of the female respondents have no relatives or friends directly participating in tourism, so the direct participation in tourism is not high. Women's social networks are narrower than men's. Under the influence of the traditional concept of "breadwinning men and homemaking women", once women lose or leave their father's or husband's family, they will break away from most of the social networks. The available social capital is very limited, and it is difficult to get rid of poverty through their own social capital. At the same time, women's information channels are blocked. Due to low education level and long-term living in relatively closed rural areas, poverty-stricken women are lack of the channel to obtain external information as well as the ability to receive external knowledge and information. Therefore, they are obviously at a disadvantage compared with men. Family economic income is limited, and financial capital is in short supply. The average yearly income of female families is 52,400 Yuan and after deducting various household consumption expenditures, most rural households make ends meet, mainly relying on borrowing money from relatives and friends to compensate for insufficient funds. The average amount of borrowings of farmers' families in Liannan County is 15,400 Yuan, and the families with borrowings account for 57.75% of the total. At present, 24.9% of the female respondents are mainly engaged in non-agricultural labor, and the average annual labor income of female employees is 56.0% of that of male employees. 8.0% of them are engaged in business, and 12.1% of them are migrant workers. Poverty-stricken women's families have difficulty in obtaining financial support and are lack of funds. 95% of women state that they have not used micro-credit loans and they received less government subsidies. Both the technical training and guidance in terms of production and the assistance in the field of medical treatment can hardly meet their actual needs. This is also an important reason for the vicious circle of women's poverty in the extremely poor areas of ethnic minorities. The acquired support is not enough, and human resource is scarce. The average scale of the family's laborers is 3.56, indicating that families have surplus labor force and have the ability of labor migration. At present, most of the women who go out to work are unmarried women with certain knowledge and skills. Since husbands work outside, 92.10% of married women take care of the elderly and children at home. About 91.49% of the female respondents have not received practical technical training. In the village, there are only 9 female cadres, which account for 2.74%. The maximum years of education of the laborers are about 9.74 on average, suggesting that a shortage of educational resources and the educational level of the family's laborers of the sample farmers are low.

4.3 Characteristics of Rural Community Inhabited by Women

In this study, 75% of the sample rural communities are poor villages, and the pure Yao villages and the Yao-Han mixed villages account for 62.5% and 37.5% of the total samples, respectively. With the continuous development of various poverty alleviation projects, the rural communities with pro-poor tourism projects account for 50% of the total samples. The support from the government for pro-poor tourism has gradually increased, and more and more funds have been invested. The average distance between various rural communities and the county is 25.99 km, and the distance between Huanglian Village and the county is 57.10 km (the longest), whereas the distance between Dazhuwan Village and the county is 5.80 km (the shortest). The average distance between rural communities and tourist attractions is 18.93 km, and the distance between Youling Village and tourist attractions is 3.7 km (the shortest), while the distance between Huanglian Village and tourist attractions is 41.30 km (the longest).

4.4 Result Analysis of the First-Level Hierarchical Linear Model

This study adopts SPSS20.0 and HLM7.03 software to conduct the mathematical analysis on the questionnaire data, and the indicators at women's individual and family levels are regressed. Through t-test, the indicators that have no significant effect on the dependent variables are eliminated, and the indicators with significant correlations are taken as the first-level indicators, while the indicators of rural communities are added as the second-level indicators. The complete two-level linear model is constructed for further analysis.

It is seen from Table 4 that the positive effect and the corresponding regression coefficient of the first-level indicators on the dependent variables in the descending order are as follows: Family's yearly income (FY: 0.172); Women's monthly income (WM: 0.168); Whether women are engaged in tourism (WT: 0.149); Women's Education (WE: 0.146); Family's laborers' education (FE: 0.139); Whether women master skills (WS: 0.137); The present value of the family's durable consumer goods (FD: 0.129); Family's cultivated land area (FC: 0.115); Family members participating in tourism (FM: 0.056). Firstly, Family's yearly income and Women's monthly income have the greatest influence, which shows that women have the most profound perception of the economic income level, and their pro-poor tourism perception is directly affected. Secondly, women engaging in tourism obtain the corresponding income from tourism, so as to improve their living standards, and they have higher pro-poor tourism perception. This is because the "nativity" of rural tourism and the "agriculture feminization" phenomenon caused by migrant working of rural men make the development, utilization and management of the resources mostly turn to the family level. For example, "living in a farmhouse, eating farmhouse food, doing farmhouse work, enjoying farmhouse fun" and other tourism projects combine the daily life, labor and tourism of rural women. Thus, women show great initiative and subjectivity in the rights and responsibilities of rural tourism resource development.

Women's Education and family's laborers' education indicate that if women have higher education levels, they will be more likely to understand and participate in pro-poor tourism. After the tuition fees for primary and secondary school students are reduced or remitted by the government, the burden of the educational expenditure on the family is alleviated, and the positive effect is remarkable. Yao women's skills are mainly scenic performance, handicraft production and embroidery, so as to increase the income of tourism products, so pro-poor tourism perception is more noticeable. Families with the larger area of cultivated land are more likely to carry out agriculture entertainment tourism, get income from tourism and reduce poverty, so as to have the positive perception of pro-poor tourism; whether family members participate

Table 4 The level-1 indicators regression results

Indicators	Coefficient	Standard error	t-Test
WA	−0.017	0.046	−0.376
WE	0.146	0.038	3.803***
WP	−0.092	0.055	−1.686*
WT	0.149	0.054	2.735***
WM	0.168	0.053	3.165***
WS	0.137	0.055	2.499**
WC	0.115	0.054	2.103**
FC	0.118	0.047	2.511**
FT	0.056	0.020	2.753***
FR	0.044	0.064	0.680
FH	−0.056	0.057	−0.984
FD	0.129	0.034	3.844***
FB	−0.032	0.048	−0.671
FY	0.172	0.065	2.657***
FS	−0.038	0.047	−0.810
FE	0.139	0.078	1.786*

(Note: $^{*}P < 0.1$; $^{**}P < 0.05$; $^{***}P < 0.01$.)

in tourism, to a certain extent, would indirectly stimulate women's ideas of further participation in tourism.

Indicators with the negative effect include whether women are engaged in agricultural production(WP: −0.092), and women in Liannan County mainly engaging in farming at home do not have time to deal with other jobs. Moreover, their education levels are not high, and the service skills and qualities of women, as tourism practitioners, cannot meet the standards, so most of the women engaging in agricultural production do not have strong awareness of participation in pro-poor tourism activities or cannot engage in the work related to tourism.

The indicators of Women's age, Family relatives and friends participating in tourism, The present value of the family's houses, Family's borrowings and The scale of the family's laborers are not markedly correlated with pro-poor tourism perception. This is mainly because the female respondents are mainly young adults, and the perception of tourism does not change with age, while fewer women's relatives and friends engage in tourism. Most women are basically reluctant to invest in tourism by selling the family's houses. At the same time, the families with larger scale of the family's laborers do not mostly participate in tourism, but they prefer to go out to work and do business, and the amount of borrowings mainly results from the imbalance of income and expenditure of farmers' families, so the effect on the local Yao women's pro-poor tourism perception is not obvious.

4.5 Result Analysis of Complete Two-Layer Hierarchical Linear Model

From the perspectives of the influencing factors of rural communities, the effect coefficient, influence and regression result significance of the same community indicator on the indicators at women's individual level are significantly different from those of the same community indicator on the indicators at women's family level. The details can be seen in Table 5.

In terms of WP-$PPTP_{ij}$ slope, with the low quality of land and the limited output of agricultural products, poor villages can obtain some national financial subsidies, but they still cannot help farmers carry out large-scale agricultural production, so women have a negative effect on the perception of agricultural development in poor villages. The development of pro-poor tourism project helps the local residents use their own agricultural products to carry out agriculture entertainment tourism, which better improves women's incomes and employment levels, so the positive effect is significant. The continuous construction of mountainous highways in rural areas has played an important role in the export and transfer of labor force, but the massive labor loss in the village has created the situation of "much land and few people" for the local agriculture. In the areas close to the center areas of the county and the tourist attractions, the sales of the local agricultural products are hammered by foreign products, resulting in the sales barriers to the local agricultural products, so

Table 5 The level-1 and level-2 indicators regression results

Indicators	Coefficient	Standard error	t-Test	Indicators	Coefficient	Standard error	t-Test
WE-$PPTP_{ij}$ slope	0.145			WP-$PPTP_{ij}$ slope	−0.148		
PV	−0.173	0.055	−3.142**	PV	−0.175	0.059	−2.951**
YV	−0.110	0.049	−2.231*	YV	−0.107	0.049	−2.175*
PP	0.133	0.055	2.416*	PP	0.138	0.057	2.418*
DC	−0.072	0.035	−2.065*	DC	−0.108	0.064	−1.680*
DA	−0.160	0.077	−2.086*	DA	−0.192	0.082	−2.336*
WT-$PPTP_{ij}$FM-$PPTP_{ij}$ 0.2290.140							
slope				slope			
PV	−0.121	0.046	−2.650**	PV	−0.160	0.071	−2.235*
YV	−0.061	0.075	−0.816	YV	−0.032	0.062	−0.523
PP	0.175	0.078	2.249*	PP	0.171	0.071	2.416*
DC	0.078	0.042	1.846	DC	−0.065	0.030	−2.161*
DA	−0.156	0.055	−2.829*	DA	−0.134	0.058	−2.305*
WS-$PPTP_{ij}$	0.181			FC-$PPTP_{ij}$	0.116		
slope				slope			
PV	−0.183	0.080	−2.278*	PV	−0.102	0.047	−2.156*
YV	−0.043	0.082	−0.523	YV	−0.004	0.052	−0.072
PP	0.171	0.071	2.416*	PP	0.126	0.050	2.545*
DC	−0.112	0.065	−1.729	DC	−0.073	0.032	−2.300*
DA	−0.180	0.081	−2.228*	DA	−0.134	0.058	−2.305*

(continued)

Table 5 (continued)

Indicators	Coefficient	Standard error	t-Test	Indicators	Coefficient	Standard error	t-Test
FM-PPTP$_{ij}$ slope	0.066			FD-PPTP$_{ij}$ slope	0.129		
PV	−0.111	0.052	−2.140*	PV	−0.124	0.053	−2.368*
YV	−0.043	0.056	−0.779	YV	−0.032	0.062	−0.523
PP	0.133	0.055	2.417*	PP	0.131	0.055	2.354*
DC	−0.082	0.039	−2.139*	DC	−0.074	0.042	−1.747
DA	−0.133	0.059	−2.243*	DA	−0.140	0.063	−2.228*
FY-PPTP$_{ij}$ slope	0.172			FE-PPTP$_{ij}$ slope	0.134		
PV	−0.121	0.051	−2.382*	PV	−0.123	0.054	−2.278*
YV	−0.042	0.057	−0.734	YV	−0.021	0.049	−0.426
PP	0.131	0.054	2.444**	PP	0.133	0.055	2.416*
DC	−0.072	0.0349	−2.065*	DC	−0.069	0.039	−1.807
DA	−0.155	0.075	−2.079*	DA	−0.167	0.076	−2.211*

(Note: *: $*P < 0.1$; $**P < 0.05$.)

the two indicators of the distance from county center and the distance from tourist attractions have a negative effect.

Regarding WT-$PPTP_{ij}$ slope, the number of women engaging in tourism in rural communities with PP is obviously more than that in rural communities without pro-poor tourism project, so the positive effect is noticeable. Pure Yao village and the distance from county center have no significant impact. Both Han women and Yao women who participate in tourism enjoy the same rights, and every village is basically connected by the road, while the traffic is more convenient, so the impact of distance is greatly reduced. The distance from tourist attractions has a negative effect, and the areas close to the scenic areas usually have a distance advantage for the local women and tourism. However, the women do not have many skills, and the training for them needs a long-term process, so in the tourist attractions, the service personnel with skills are mostly employed from other areas. Therefore, at present, the local women's perceptions are mostly negative.

Concerning WM-$PPTP_{ij}$ slope, the monthly income of the poor villages has a negative impact on women's perceptions, because most women in poor villages do not obtain poverty alleviation subsidies directly from the government. The indicator of pro-poor tourism project has a marked positive effect, and compared with women who do not participate in tourism, the monthly incomes of women who engage in tourism increase more rapidly. Pure Yao village has no significant influence, because there is no obvious ethnic discrimination and difference in women's employment opportunities, and both Yao women and Han women are equal in employment. The distance from county center and the distance from tourist attractions have a negative effect, and with regard to the women close to the scenic areas and the county, the increase in costs of living, transportation and accommodation and the environmental damage caused by tourism lead to more serious living and environmental burdens on the local residents.

In terms of WS-$PPTP_{ij}$ slope, pro-poor tourism project has carried out the relevant skill training for the local residents to some extent, which is helpful and makes the women's perceptions have a positive effect. However, the skill training for women is insufficient in the poor villages, and at the same time, the traditional embroidery techniques of Yao nationality can be sustainably inherited and developed through a long-term process, so women's perceptions of Poor village have a negative effect. Pure Yao village and the distance from county center have no significant effect, and the distance from tourist attractions has a noticeable negative impact. In the rural communities close to the scenic areas, the Yao farmers have very limited social network and few opportunities to obtain social resources and learn skills, and most of them sell their own agricultural products and small commodities in the scenic areas but do not have more technical breakthroughs.

Regarding FC-$PPTP_{ij}$ slope, Poor village has a negative impact on women's perceptions, and this is mainly due to the poor quality of land and the low output of agricultural products in poor villages. pro-poor tourism project has a marked positive effect, and the family's cultivate land area is large, while the left-behind women at home have more room for carrying out agriculture entertainment tourism projects, so as to obtain more benefits. The distance from county center and the distance from tourist attractions have a remarkable negative effect, if the distance from the scenic areas and the county is shorter, the freight and commuting costs for sales of agricultural products and tourism products in the county and the scenic areas will be greatly reduced. However, female family members in rural communities close to the county are more inclined to go out to work and earn more money. Since the investment in the agriculture entertainment tourism projects requires the huge early-stage investment, the women close to the scenic areas are more inclined to sell small commodities in the scenic areas, resulting in abandoned and idle land at home. Thus, the negative effect is obvious.

Concerning FM-$PPTP_{ij}$ slope, Pure Yao village has no significant impact, and pro-poor tourism project (PP) has a noticeable positive effect, whereas other factors have a negative influence. Among them, the average scale of family members participating in tourism in poor villages in Liannan County is about 1–2 persons, and more family members choose to engage in agricultural production and go out to work, while most of the rural households do not meet the standards of poor households stipulated by the government, so the relevant subsidies for poor villages have not been actually obtained. Therefore, the negative effect is significant. The distance from county center and the distance from tourist attractions have a remarkable negative effect. If family members are engaged in tourism, and the distance from the county and the scenic areas are shorter, the employment opportunities will be more diversified, and livelihood selections will tend to be more diversified as well, so they are not limited to engaging in tourism. Hence, women's individual perceptions have a negative impact.

With regard to FD-$PPTP_{ij}$ slope, Poor village has a negative effect on women's perceptions, and poor villages are characterized by the lack of physical assets, the low value of the family's durable consumer goods, the poor liquidity and the low effect on reducing livelihood risks. The distance from tourist attractions has a negative impact, and if the distance from the scenic areas is shorter, the purchase prices of various durable products will be higher, and the economic burdens on the family will be greater, while the risks of poverty will be increased. PP has a marked positive effect, whereas Pure Yao village and the distance from county center have no significant impact.

In terms of FY-$PPTP_{ij}$ slope, Poor village has a negative effect on women's perceptions, because compared with non-poor villages, the yearly income of rural households in poor villages is lower, and the vulnerability of rural households in poor villages is higher. pro-poor tourism project has a significant positive effect on the increase in the yearly income, while Pure Yao village has no obvious impact on women's perceptions and yearly incomes. If the distance from the county and the scenic areas are shorter, the family income will increase, but the rents and land prices will increase as well due to tourism. Therefore, the living costs of the local residents will be higher, and the gap between the rich and the poor will be more obvious. The distance from county center and the distance from tourist attractions have a negative influence on women's perceptions.

Regarding FE-$PPTP_{ij}$ slope, Pure Yao village and the distance from county center have no noticeable influence. There is no ethnic difference in labor education, and both Yao residents and Han residents can receive education and training. At present, the roads connecting the villages are constructed perfectly, which is convenient for the family's laborers to go to the county to receive the related education and training. pro-poor tourism project has a remarkable positive effect, and pro-poor tourism project not only stimulates the employment of the local laborers, but also improves the educational levels of the local laborers. Therefore, the positive effect is obvious, indicating that the poverty alleviation projects of the Chinese government focus on the education of laborers and the training of professional abilities. The distance from tourist attractions has a negative effect. Generally speaking, if the distance from the scenic areas is shorter, the educational conditions of the family's laborers should be better under the influence of the economic radiation of the scenic areas. However, women's perceptions have a negative impact, which reflects that in remote areas, more family's laborers go out to work, and these laborers directly working in the scenic areas are more likely to obtain favorable conditions for education and professional training.

5 Analysis of Rural Community Indicators on Pro-Poor Tourism Perception

According to the original variances and the conditional variances of the indictors listed in Table 6, the corresponding explanatory variances are calculated 0.5 is taken as the standard line, and it is seen that the indicators at women's individual and family levels that have a greater influence on pro-poor tourism perception include whether women master skills, women's Education, whether women are engaged in tourism, family's cultivated land area and family's yearly income. The effects of other variables on pro-poor tourism perception are attributed to the indicators at the rural community level, including whether women are engaged in agricultural production, women's monthly income, family members participating in tourism, the present value of the family's durable consumer goods, and family's laborers' education. Rural communities provide the local women who are engaged in agricultural production with cash subsidies for crop planting, cash crop planting, livestock breeding and other related agricultural production activities, which greatly supports the development of the local agriculture. Rural communities use pro-poor tourism project to increase women's employment opportunities and women's monthly incomes. The increase in the monthly income has enhanced women's desire to purchase the family's durable consumer goods, effectively reduced the poverty of the local women, decreased the rate of women's return to poverty and improved the overall living standards of farmers. At the same time, it has strengthened women's pro-poor tourism perception and made full use of the status and role of women, as the actual household heads in the families. This has also led to more opportunities for other family members to participate in tourism. If more family members participate in tourism, the local farmers' perceptions will be higher. Family's laborers' education is the main way to accumulate human capital of farmers' families, and education expenditure accounts

Table 6 Indicators original variance, conditional variance and explained variance

Indicators	Original variance	Conditional variance	Explained variance
WE	0.0176	0.0031	0.82
WP	0.0199	0.0082	0.59
WT	0.0173	0.0050	0.71
WM	0.0112	0.0057	0.49
WS	0.0178	0.0023	0.87
FC	0.0116	0.0040	0.65
FM	0.0118	0.0063	0.47
FD	0.0154	0.0072	0.53
FY	0.0172	0.0066	0.62
FE	0.0212	0.0129	0.39

for a larger proportion of the family's daily expenditure. Rural communities use pro-poor tourism project to support the education and training of the family's laborers and alleviate the burden on farmers. Among the female samples of Yao nationality in Liannan County, women who go out to work have higher educational levels. It can be seen that the support from rural communities plays a significant role in women's pro-poor tourism perception. Therefore, it is necessary to strengthen the construction of the local rural communities and formulate and perfect the relevant policies and regulations.

6 Conclusion and Discussion

In order to more accurately extract the key influencing factors of women's pro-poor tourism perception in the ethnic minority areas of northern Guangdong Province in China. As indicated from the results, the key factors at women's individual and family levels affecting pro-poor tourism perception include Women's monthly income, Family members participating in tourism and Family's laborers' education, all the indicators have a positive effect. The key factors at the rural community level affecting pro-poor tourism perception include Poor village, pro-poor tourism project, Pure Yao village, the distance from tourist attractions and the distance from county center, and wherein, the indicator of pro-poor tourism project has a positive effect, while other indicators have a negative effect. Based on the analysis results of the above key influencing factors, the suggestions for improving the positive effect of pro-poor tourism perception are put forward as follows.

The overall improvement of poverty-stricken women's livelihood capital is taken as the direction of the effort to fight against women's poverty and promote women's development, which also means pursuing women's development goals from more dimensions, including achieving the combination of income increase and overall development, the combination of gender equality and power construction and the combination of individual ability improvement and overall environment improvement. The support from rural communities is very significant. In terms of community policy, China's anti-poverty is dominated by the government. With the promotion of pro-poor tourism project and the support of rural revitalization strategy, firstly, there is a need to make full use of the effective funds and various resources from government departments and non-government departments, actively increase the investment in educational funds and improve the educational levels of women and the family's laborers.

Meanwhile, it is necessary to adopt rural pro-poor tourism project to actively stimulate women's employment, achieve rural women's poverty alleviation and increase women's individual and family incomes. Moreover, on the one hand, the families' direct participation in rural tourism management, such as the operation of agriculture entertainment tourism projects and the management of rural family homestays, should be increased, so that women can continue to engage in agricultural production at home, take into account housework and obtain benefits from tourism, so as to improve the effect of pro-poor tourism; on the other hand, rural women should be encouraged to actively participate in rural tourism reception services and work in the scenic areas, and their own agricultural and sideline products should be sold through the development of rural tourism, so as to expand the sales channels of agricultural products and increase the sales prices of agricultural products.

In addition, there is a need to obtain rents through participation in rural tourism cooperatives and land transfer, get dividends through the participation of capital, manpower and land in rural tourism management and further protect and inherit the Yao culture through the training of women's Yao culture performance and embroidery skills. Compared with government organizations, non-governmental organizations (NGOs) have advantages in value orientation, cost control, closeness to the people, flexible innovation, etc., which makes them play an active role in alleviating women's poverty, especially humanistic poverty, such as promoting women's employment, establishing gender equality mechanism, safeguarding poverty-stricken women's human rights, promoting poverty-stricken women's participation, increasing social capital and promoting social harmony. In terms of community transportation, since the distance from county centerand and the distance from tourist attractions directly affect women's pro-poor tourism perception, women get income by working in the county and scenic areas, but the effect of perception is negative. It is suggested that traffic accessibility is the core element of pro-poor tourism in the Yao area of northern Guangdong Province, and the convenient, fast and safe tourism road networks should be formed by accelerating the construction of the connecting lines between expressways and scenic areas, especially the construction of transportation road networks between different scenic areas. It is necessary to take the county and scenic areas as the core areas to accelerate the construction of various types of catering, accommodation and commercial supporting facilities, so that the local female employees could effectively enjoy the benefits brought by pro-poor tourism, and the negative effect of their perceptions could be transformed into the positive effect.

As the division of women's roles has its own vulnerability, as the novel coronavirus pandemic sweeps the world, women's poverty is exacerbated, so in the future it is necessary to make full use of women's important roles in the protection of national cultural resources and the inheritance and development of national traditional handicrafts. At the same time, improving the survival skills of Nationalities of China women and coping with the risk of unemployment caused by covid-19. Besides, there is a need to establish the pro-poor tourism-based benefit sharing-centered community participation mechanism to guarantee the interests of poor women. By enhancing women's rights of speech and benefit sharing in tourism development and emphasizing the importance of women in development, women are shifted from passive subordination to active action, and unequal power relations are broken, so that they can play a greater role in the policy-making process of resource allocation and gain more decision-making rights in the development process. Only in this way can women have real development opportunities.

Acknowledgements We thank the funding support from the National Natural Science Foundation of China -Youth Project (Grant NO. 42101310):Research on the Measurement and Driving Mechanism of relative poverty of rural residents in the Typical Mountainous Areas of Southwestern China and The characteristic innovation project of the Department of Education of Guangdong Province, China (Grant NO. 2021WTSCX029):Study on spatial differentiation and high-quality development path of guangdong-hong Kong-macao Bay area vacation leasing under sharing economy. The contents of this paper are solely the responsibility of the authors and do not represent the official views of the a forementioned institutes and funding agencies.

References

1. Churchill, S, A., Smyth, R.: Ethnic diversity and poverty. World. Dev. 95, 285–302 (2017)
2. Narayan, A., Van der Weide, R., Cojocaru, A., et al.: Fair Progress? World Bank Publications, Economic Mobility Across Generations Around the World (2018)
3. Sofield, T., Bauer, J., De Lacy, T., Lipman, G., Daugherty, S.: Sustainable tourism-eliminating poverty (ST-EP) (2004)
4. Mason, A, D., Badiani, R., Van Nguyen, T., Patrick, K., Del Carpio, X.: Toward gender equality in East Asia and the Pacific. A companion to the world development report. World Bank East Asia and Pacific regional report. World Bank, Washington, DC (2012)
5. Xiang, M., Li, Y.: Incorporating Gender Perspective into pro-poor tourism: a reflection on Guangxi's Ethnic pro-poor tourism. J. Guangxi Econ. Manag. Gadre Coll. **1**, 102–107 (2015)
6. Pearce, D.: The feminization of poverty: women, work, and welfare. Urban Soc. Chang. Rev. 11(1), pp. 28–36 (1978)
7. Goldberg, G, S., Kremen, E.: The feminization of poverty: only in America? Only in America? (No. 117). ABC-CLIO (1990)
8. Bird, K., Shinyekwa, I.: Even the 'Rich' are vulnerable: multiple shocks and downward mobility in rural Uganda. Dev. Policy Rev. **23**, 55–85 (2005)
9. Razavi, S.: World development report 2012: gender equality and development—a commentary. Dev. Chang. **43**, 423–437 (2012)
10. Liu, J., Wang, X.: Poverty reduction effect of female marriage in poor areas. J. South China Agric. Univ. (Soc. Sci. Ed.). **4**, 106–118 (2018)
11. Budowski, M., Tillmann, R., Bergman, M, M.: Poverty, Stratification, and gender in Switzerland. Schweiz. Z. Soziol. 28, 297–318 (2002)

12. Moghadam, V, M.: Globalizing women: transnational feminist networks. JHU Press (2005)
13. Bergmann, B.R.: The economic emergence of women, p. 58. Palgrave Macmillan, New York (2005)
14. Organisation for Economic Co-operation and Development, & Organisation for Economic Co-operation and Development Staff OECD employment outlook 2011. OECD Publishing, Paris (2011)
15. Gökovalı, U.: Everyone's own poverty: gendering poverty alleviation policies in Turkey. In: Women's Studies International Forum. Vol. 41, pp. 65–75. Pergamon (2013)
16. Klasen, S., Lechtenfeld, T., Povel, F.: A feminization of vulnerability? Female headship, poverty, and vulnerability in Thailand and Vietnam. World Dev. **71**, 36–53 (2015)
17. Goldberg, G, S.: Poor women in rich countries: the feminization of poverty over the life course. Oxford University Press (2010)
18. Van de Walle, D.: Lasting welfare effects of widowhood in Mali. World Dev. **51**, 1–19 (2013)
19. Chantarat, S., Barrett, C, B.: Social network capital, economic mobility and poverty traps. J. Econ. Inequal. 10, 299–342 (2012)
20. Kabir, A., Maitrot, M, R, L.: Exploring the effects of health shocks on anti-poverty interventions: experience of poor beneficiary households in Bangladesh. Cogent Med. 5, 1468233 (2018)
21. National Bureau of Statistics of China.: China statistics yearbook. http://www.stats.gov.cn/tjsj/ndsj/2019/indexch.htm
22. Deng, X., Zeng, L., Luo, M.: Study on the accurate identification of tourism poverty alleviation under the background of accurate poverty alleviation. Ecol. Econ. 31, 94–98 (2015) Ecol.
23. Millar, J., Glendinning, C.: It all really starts in the family': gender divisions and poverty. women and poverty in Britain: the 1990s, Hemel Hempstead, Harvester Wheatsheaf (1992)
24. Liu, X.: Review of the literature on women's poverty in China in the past 40 years. Collection of Women's Studies **1**, 116–123 (2015)
25. Ashley, C., Roe, D.: Pro-poor tourism strategies: making tourism work for the poor: a review of experience. (2001)
26. United Nations World Tourism Organization (UNWTO).: global report on women in tourism: preliminary findings (2010)
27. Fan, X.: Female employees in overseas tourism. collection of women's studies, 5, 105–112 (In Chinese) (2012)
28. Ferguson, L, Alarcón, D, M.: Gender and sustainable tourism: reflections on theory and practice. J. Sustain. Tour. 23, 401–416 (2015)
29. Harrill, R.: Residents' attitudes toward tourism development: a literature review with implications for tourism planning. J. Plan. Lit. **18**, 251–266 (2004)
30. Marin, V., Palmisani, F., Ivaldi, R., Dursi, R., Fabiano, M.: Users' perception analysis for sustainable beach management in Italy. Ocean Coastal. Manag. **52**, 268–277 (2009)
31. Sainsbury, D., Morissens, A.: Sweden: the feminization of poverty?. Poor women in rich countries: the feminization of poverty over the life course, 28 (2010)
32. Kwaramba, H., M, Lovett, J, C., Louw, L., Chipumuro, J.: Emotional confidence levels and success of tourism development for poverty reduction: the South African Kwam eMakana home-stay project. Tour. Manag. 33, 885–894 (2012)
33. Von der Weppen, J., Cochrane, J.: Social enterprises in tourism: an exploratory study of operational models and success factors. J. Sustain. Tour. **20**, 497–511 (2012)
34. Duffy, L, N., Kline, C, S., Mowatt, R, A., Chancellor, H, C.: Women in tourism: shifting gender ideology in the DR. Ann. Tour. Res. 52, 72–86 (2015)
35. Khatiwada, L, K., Silva, J, A.: Mitigating gender inequality in rural regions: the effects of tourism employment in Namibia. Int. J. Tour. Res. 17, 442–450 (2015)
36. Jaggar, A, M.: Is globalization good for women? Comp. Lit. 53, 298–314 (2001)
37. Job, H., Paesler, F.: Links between nature-based tourism, protected areas, poverty alleviation and crises—the example of Wasini Island (Kenya). J. Outdoor Recreat. Tour. **1**, 18–28 (2013)
38. Soliman, M, S.: Pro-poor tourism in protected areas–opportunities and challenges: "The Aase of Fayoum, Egypt". Anatolia. 26, 61–72 (2015)

39. Guo, S.: Industrial chain tracing method for assessing pro-poor tourism impact. Tourism Tribune **30**, 31–39 (2015)
40. Li, Y., Zou, M., Sun, X.: The research frontier and evolution of domestic tourism poverty alleviation——an analysis of knowledge mapping based on cite space. J. Baoding Univ. **32**, 46–54 (2019)
41. Goodwin, H.: Tourism, local economic development, and poverty reduction. Appl. Res. Econ. Develop. **5**, 55–64 (2008)

Analysis of Tourist Storytelling Behavior and Revisit Intention in Red Tourism by Flow Experience and Patriotism

Dengming Xie

Abstract Red tourism in China is the visit to sites significant to the Chines revolution history. Red tourism plays the role of patriotism education by influencing tourists' cognition construction and reconstruction. In this sense, the inheritance of red culture embedded in red tourism destinations depends not only on the promotion of suppliers but also on tourist participation. However, how the tourism experience leads to tourists' cognitive construction or reconstruction and further affects their inheritance behavior of red culture is an important and unanswered question. Thus, this study attempts to investigate the effects of flow experience and patriotism on tourist storytelling behavior and revisit intention. The PLS-SEM was employed to analyze data collected from 208 tourists. The results indicate that flow experience positively affects patriotism, storytelling behavior, and revisit intention and that patriotism mediates the effects of flow experience on both storytelling behavior and revisit intention. This study contributes to our understanding of tourist patriotism and their inheritance behavior of red culture.

Keywords Red tourism · Flow experience · Patriotism · Storytelling behavior · Revisit intention

1 Introduction

Tourism experience is a continual process that has potentially profound and long-term impacts on tourists from the during-trip to the post-trip stage [1, 2]. Recently, scholars have examined that the experience of visiting memorials, museums, and other nationalistic historic sites reinforces tourists' cognition toward a nation/country, such as national/country/party identity [3–5]. Specifically, Zhao and Timothy [2] stated that red tourism in China reflects a role of patriotism education by influencing tourists' cognition construction and reconstruction [4]. Although previous research has shed

D. Xie (✉)
Macau University of Science and Technology, Avenida Wai Long, Taipa, Macau, People's Republic of China
e-mail: dengmingxie@outlook.com

Y. Luo et al. (eds.), *Tourist Experience and Perception in the Pandemic and Post-Pandemic Era*, https://doi.org/10.1007/978-981-96-1820-0_12

light on the potential association between tourism experience and cognition construction and reconstruction in the context of red tourism [5], scholars have not yet illustrated the holistic influence mechanism of how tourism experience shapes tourists' cognition and ultimately affects their behavioral tendency. Indeed, tourist cognition changes and favorable post-trip behavior are meaningful for the development of red tourism in China. Thus, there is a need to focus on the above issue.

Existing qualitative literature has addressed that tourism activities in a specific red tourism destination led domestic tourists to build and elevate patriotism [6]. Patriotism is a key term related to the practical development and academic research of red tourism in China [4], but our understanding of the patriotism aroused by red tourism activities is relatively limited. Meanwhile, given that red tourism in China shoulders the responsibility of patriotism education [4, 5], this study operationalizes the process of tourist cognition construction and reconstruction within red tourism activities into patriotism.

Nowadays, tourists are actively involved in the tourism industry and are seen as value co-creators, creating value for themselves, tourism destinations, and even the industry [7]. In other words, tourists play a nonnegligible role in destination marketing promotion. As such, the inheritance of red culture depends not only on the storytelling of red tourism destinations from the perspective of suppliers but also on tourists' role in storytelling and revisit behavior. However, the knowledge about tourists' co-creator role in storytelling is limited and the whole picture of its' generation mechanism beginning with the tourism experience is unclear, whereas this topic is important for red tourism development, thereby, exploring storytelling and revisit behavior of red tourists is a valuable attempt.

In sum, how flow experience in red tourism leads to tourists' cognitive changes and further affects their red culture inheritance behavior is an important and unanswered question. Drawing on the above discussion, this study attempts to investigate the effects of flow experience and patriotism on tourist storytelling behavior and revisit intention. The PLS-SEM was employed to conduct data analysis procedure with 208 valid data from red tourists. This study contributes to our understanding of tourist patriotism and their inheritance behavior of red culture.

2 Literature Review

2.1 Red Tourism

Red tourism in China was officially proposed in 2004 by the central government department [5], and nationwide strategies have then been implemented to facilitate the development of red tourism [4]. Scholars have continuously refined the definition of red tourism for corresponding to its practical development, and the mainstream view is of red tourism refers to a kind of theme tourism activity that organizes and receives tourists for memorial study and visit, it takes the memorial sites and symbols with patriotic characteristics that occurred on the land of China since 1840, especially during the period of the Revolutionary War (1921–1949) under the leadership of the Communist Party of China, as the vectors, with historical events, historical personage, culture and spirit as the connotations [8].

Previously, red tourism is considered a nationwide political socialization campaign for sustaining communist/country identity in which government plays the main role from both top-down and bottom-up [9]. However, red tourism nowadays is more interactive, experiential, entertaining, and educational, as a result, people are attracted by this kind of special interest tourism and are more likely to visit red tourism destinations willingly [4, 6, 10, 11]. So red tourism is moving towards a more market-oriented development direction, different from the previous government-oriented situation [12]. As such, academic research has also gradually shifted from a government-led/supplier-led perspective to a tourist-led perspective. For example, the recent work by Tang et al. [4] focused on how tourist perception of the cultural atmosphere constructed by the red tourism destination affects their subsequent tourism experience and red memory. Research also determined that satisfaction positively affects tourist intention to visit similar red tourism destinations [13, 14].

However, academic attention to red tourists' cognition, emotion, and behavior is fledgling and insufficient. Conversely, the real-world phenomenon that tourists share their tourism experience related to red culture (e.g., "We were born under the five-star red flag and grew up in the spring breeze") on various social networking sites (e.g., WeChat moments, Douyin, etc.) is substantial. And, that phenomenon is the so-called 'storytelling', indicating that tourists are playing the role in the inheritance of red culture embedded in red tourism destinations. Thus, drawing on literature and the above phenomenon, the current study attempts to look closer at tourist responses toward the tourism experience of red tourism, that is patriotism, storytelling behavior, and revisit intention.

2.2 *Patriotism in Tourism*

Since the end of World War II, the establishment of new countries around the world has increased significantly, accordingly, related issues such as patriotism and nationalism attracted widespread attention at that time [15–17]. Lately, patriotism is understood as the attitude that can influence individual behavior [18, 19]. Particularly, patriotism is an individual's attitude toward the country [20], specifically the individual's love and respect for the country and its symbols [17, 21]. Patriotism arises when personal values/beliefs are aligned with national ones, and patriotism is derived from cultural heritages, traditions, and national symbols that embed national values/beliefs [21]. In this sense, cultural heritage tourism is treated as a path to establish patriotism in the minds of tourists [3].

In the tourism literature, only a few research has empirically identified the relationship between patriotism and behavioral components. For instance, Qu et al. [21] determined that tourist patriotism positively affects their psychological ownership toward the country and civilized tourism behavioral intention when traveling abroad. Choe et al. [19] identified the effects of patriotism on mega-sporting event attendance intention (as an audience in the Olympic event) and related media consumption intention. After reviewing the literature, we conclude that there is a trace that the term patriotism always emerges in the research context associated with collective focuses, such as international travel, the Olympic games, national policy, immigration issues, and so on [22–24]. In the context of red tourism, although the inevitable connection between the red tourism experience and tourist patriotism is widely accepted, scant research has been conducted on the antecedents and consequences of patriotism. Given that "red tourism has also been heavily leveraged as a tool for patriotic education" [4], this study takes patriotism to understand tourist post-trip storytelling behavior and revisit intention.

2.3 Flow Experience

The flow theory, put forth by Csikszentmihalyi [25], refers to a subjective optimal experience, or an ecstasy state, lived by individuals during events or tasks while being performed [26], which feature high levels of concentration, complete immersion in experiences or activities, and the absence of awareness of time and space [27]. State of flow is regarded as central of the theoretical and operational framework of the tourism experience, as tourists, who move away from usual environment towards unusual environment, are primarily attracted and motivated by on-site experiences that immerse them in rich and positive psychological states such as delight, enjoyment, and internal harmony [28].

The way in which tourists assess the optimal experiences, and the underlying reasons for such experiences have sparked research interest in academia [29]. The difficulty in operating and evaluating flow experience, causing by the confusion of antecedents, core elements, or consequences of flow, is the most common criticisms of this concept [28]. Some researchers argued that the flow experience is like a continuous slide upwards from a normal to an optimal psychological state [30]. Nevertheless, "what scholars aim to measure or are able to measure (through a scale) is the static optimal level of experience" [31]. Thus, flow experience is usually measured as a subjective optimal experience and by several items that describe the features of flow state, such as totally focus on, high level of immersion and lost track of time [32].

In terms of the underlying reasons for flow experience in tourism context, according to Csikszentmihalyi [25], flow is a state that emerges when challenges and skills reach a balance. More specifically, individuals experience flow state when they voluntarily participate in events or activities, and feel in harmony with the environment, and their skills match the challenges [33]. In line with this logic, Coble et al. [34] stated that flow experience occurs when the objective and perceived challenges of hiking strike a balance with the skills of hikers. Similarly, Wu and Liang [33] found that tourist rafting skill, challenge of rafting, and playfulness positively affect flow experience. As research deepens, the applicability of flow concept has been extended to various types of consumer activities. Prior studies have determined that personality, intrinsic, and extrinsic motivation are the first-tier drivers of flow experience [28], while related cognitive states such as engagement, involvement, and enjoyment are the second-tier drivers of flow experience [31].

Tourists who participate in red tourism activities often have an inherent motivation to study history and appreciate red spirit, meantime, the scenarios of red tourism are often independent from the outside world and can awaken tourists' strong emotions such as admiration, sad, pride [4, 13]. Thus, the intrinsic and extrinsic motivation are substantial and can stimulate tourists to immerse themselves in the red atmosphere and narrative of red tourism destinations, creating a flow experience.

2.4 Storytelling Behavior

Storytelling is a core feature of interaction and communication in daily life [35]. Storytelling is not the behavior of a specific group or character, thus, the definition of storytelling depends on the context and character's perspective [36]. Theoretically, storytelling is of a communication process whereby people share information and experiences through the mode of story and narration [37]. Storytelling is usually told with a purpose in mind, in this sense, storytelling is only a method of communication to approach certain purposes. According to Keskin et al. [38], storytelling refers to share information and experiences through narrative and anecdotes to communicate complex ideas, concepts, and causal connections, and then build connections and associations.

When interpreting storytelling in marketing context, it is imbued with the connotation of adverting and persuading, and emphasizes transmitting information about a brand, a product, or service in a story-like format [39]. Similarly, in the field of tourism marketing, storytelling is also treated as a strategic resource in the tourism industry [36]. Especially the storytelling behavior of tourists in the post-travel stage has the most commercial value, as it integrates emotions, fantasies, and thoughts and thus has stronger persuasive effect than general positive word-of-mouth [36].

As an experience-based industry, tourism industry exposes tourists in the extraordinary experience that form the basis for storytelling. Extraordinary experience can stimulate tourists to share it with others via online or offline channels at the during-travel and post-travel stages. Flow experience, representing optimal state during a trip, is therefore (re)presented, (re)produced, and (re)created through narratives on postcards, photographs, and through storytelling [40]. As such, tourism experiences become a new information source through storytelling. Broadly speaking, storytelling not only strengthens the narrator's travel memory and establishes a stronger connection with the destination, but also enables the information of the destination to be received by others. Therefore, from the practical perspective, storytelling around a red tourism destination is a red cultural inheritance behavior with important practical implication. Unfortunately, to date, storytelling is generally an under-researched theme in the topic of red tourism. Therefore, there is a need to determine its drivers to promote the development of red tourism and the inheritance of red cultural.

2.5 *Revisit Intention*

Behavioral intention is about anticipated or planned future behavior, which refers to the intensity of an individual's spontaneous plan to implement a specific behavior, and can also be understood as the effort degree of an individual to execute a specific behavior [41]. Fishbein and Ajzen proposed the theory of rational behavior in 1975, and Ajzen subsequently proposed the theory of planned behavior in 1991, believing that behavioral intention is a necessary process or a prerequisite for actual behavior. Soon, this theory was introduced into tourism research [41]. Given that behavioral intention has effective predictive and explanatory power on actual behavior, and considers the operability, in the field of consumer behavior research, scholars often utilize consumer behavioral intention to operationalize consumer actual behavior [42].

Revisit intention is one of behavioral intentions and a kind of post-consumption behavioral intention, which refers to the likelihood of a tourist repeating a tourism activity that he or she has experienced before or revisiting a tourism destination that he or she has been to before [43]. Comparted with first-time visitors, repeat visitors means lower marketing costs, higher satisfaction levels, and positive word-of-mouth [44]. Therefore, scholars and practitioners in the field of tourism are interested in understanding why tourists would like to revisit the same or similar destination [45]. The decision-making process of choosing a travel destination is complex and impacted by multiple factors. The author argues that revisiting the same or similar red tourism destination is a typical inheritance behavior of red culture, thus, exploring the antecedents of revisit intention of red tourists is of theoretical and practical contribution.

3 Hypothesis Development

3.1 *Flow Experience and Its Consequences*

As noted by [30], the concept of flow experience seems to be undervalued in tourism research because tourism activities can easily make people in an optimal mental state [28]. 'Flow' can be described as the subjective experience that individual experiences during an event or task being performed, usually associated with joy, enjoyment, and loss of control, which is, in a way, the intended outcome of a travel experience [28].

Applying the idea of flow theory in the context of red tourism, tourists receive and process emotional and cognitive stimulation from materials such as life stories of historical people, pictures, and videos [28, 30], and that procedure makes them immerse in the red cultural atmosphere constructed by the destination, as such, flow experience occurs. In other words, the interactions between tourists and the red cultural atmosphere can lead to a special mental state, which is helpful to create optimal flow experiences [30]. Furthermore, that mental state might exert powerful effects on tourists' emotional, cognitive, and even behavioral responses to a given object [46], that is, tourist patriotism, storytelling behavior, and revisit intention in this study.

Rutherdale [47] stated that "the frequent occurrence of flow state in liminal situations, particularly those where communities emerged with pronounced intensity". In line with that statement, tourists are more likely to place themselves in the broad community, named a party/nation/country, when they immerse in the red cultural atmosphere which bridges them with a party/nation/country, which in turn results in the construct and development of patriotism. After that, tourists tend to revisit the same or other red destinations, as well as to say good things about their tourism experience and red destinations online or in the real world. Thus, it is reasonable to assume that the flow experience in red tourism destinations promotes tourist patriotism, storytelling behavior, and revisit intention. Meanwhile, previous studies have well examined that the flow experience positively promotes the formation of favorable post-trip behaviors and behavioral intentions, involving satisfaction, destination loyalty, WOM, and storytelling [40, 28]. Thus, the following hypotheses were proposed.

H1. Flow experience positively affects tourist storytelling behavior.

H2. Flow experience positively affects tourist revisit intention.

H3. Flow experience positively affects tourist patriotism.

3.2 *Patriotism, Storytelling Behavior, and Revisit Intention*

In addition, as suggested by Zhao and Timothy [2], tourism alone does not guarantee cognition change, and it can successfully lead to cognition (re)construction also relies on contradictions, personal preferences, beliefs, values, encounters, and many other elements. Cognition (re)construction is mainly reflected in the adjustment of preconceptions [2]. The findings of Tang et al.'s [4] work indicate that tourists develop, improve, and correct their red memory by participating in red tourism, which is consistent with the above opinion. Throughout this cognition (re)construction procedure, tourists co-creating their personal and unique tourism experience, and that co-creating process continues in the post-trip stage in which tourists consolidate and show their connection and consistency with the red destination (or the party/nation/country) by storytelling and revesting [7]. Therefore, we assume that patriotism can facilitate storytelling behavior and revisit intention. Thus, the following hypotheses were proposed:

H4. Patriotism positively affects tourist storytelling behavior.

H5. Patriotism positively affects tourist revisit intention.

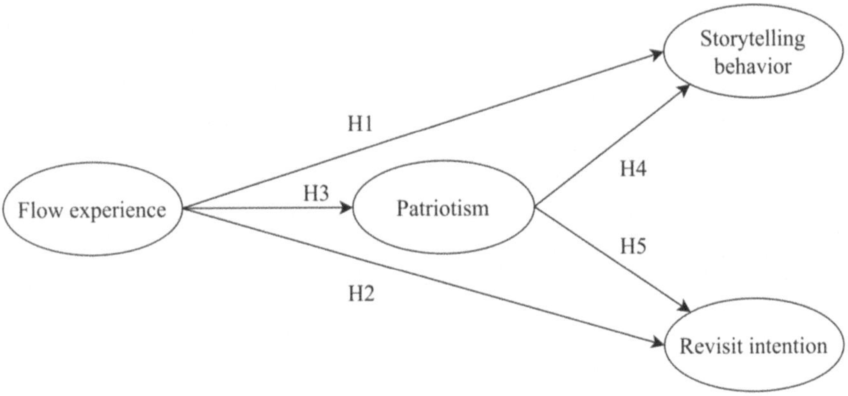

Fig. 1 Research model

In sum, the research model presenting the above relationships is shown in Fig. 1.

4 Methodology

4.1 Measures and Questionnaire

As shown in Table 1, the measurable items for flow experience were adopted from Ding and Hung [48] and Tang et al. [30]. The three items for evaluating tourist patriotism aroused by their travel to red tourism destinations were modified from the research by Sidanius et al. [17]. The scale of storytelling behavior was revised by the work of Zhong et al. [6]. The scale of intention to visit the same or other red tourism destinations was adopted from Wong et al. [49] and Meng et al. [50].

In the first part of the questionnaire, we presented introductory messages and voluntary participation principles of the current survey. Then, a screening question was set to approach the target respondents, thus, those who answered "yes" to the question "Have you been to red tourism heritage destinations in the past six months" were invited to fill out the questionnaire. After that, respondents rated all 16 questions on a 7-point Likert scale (1 = strongly disagree, 7 = strongly agree). In the last part, the information on respondents' demographics was collected. The back-translation procedure proposed by Brislin [51] was conducted by 3 experts in tourism and hospitality to check for translation biases and to ensure the content validity of the questionnaire.

Table 1 Items and references

Construct and item	Reference
Flow experience (FE)	Ding and Hung [48]; Tang et al. [30]
I thought I have experienced flow during visiting the red destination	
In general, I frequently experienced flow when in the red destination	
Most of the time when in the red destination I felt that I am in a flow state	
Visiting the red destination excites my curiosity	
Visiting the red destination arouses my imagination	
Patriotism (PTT)	Sidanius et al. [17]
I have great love for my country	
I am proud to be a Chinese	
I feel much affection for China	
Storytelling behavior (STB)	Zhong et al. [6]
I posted photos on social media to share my vacation stories	
I talked about my vacation with others by telling stories	
I provided a narrative of my vacation experiences to family and friends	
I shared photos of my vacation experiences with others	
Revisit intention (RI)	Wong et al. [49]; Meng et al. [50]
I would visit other red destinations in the future	
If given the opportunity, I would like to have another red tourism	
I am willing to re-participate red tourism in the near future	
I am planning to re-participate in red tourism in the near future	

4.2 Data Collection

A pilot survey with 20 students majoring in tourism and hospitality was carried out in November 2021 to assure the accuracy and readability of all the questions. After receiving positive feedback, the formal survey was conducted in December 2021. Finally, 260 questionnaires were collected through an online survey platform (www.wjx.cn) [52], of which 208 were valid after eliminating invalid data, for a valid response rate of 80.0%. The sample size was more than ten times the number of indicators in this study, so the dataset was appropriate for structural equational modeling (SEM) analysis [53]. As shown in Table 2, among the 208 valid respondents, females accounted for 60.1% while males accounted for 39.9%. The majority

Table 2 Demographic characteristics of respondents (N = 208)

Project	Category	Frequency	Percentage (%)
Gender	Male	83	39.9
	Female	125	60.1
Age	18–20	11	5.3
	21–25	87	41.8
	26–30	73	35.1
	31–35	34	16.3
	36 and above	3	1.5
Education	Senior secondary and below	43	20.7
	Undergraduate	161	77.4
	Postgraduate	4	1.9
Career	Student	97	46.6
	Enterprise staff	51	24.5
	Freelancers	24	11.5
	Others	36	4.3
Monthly income	3000 and below	75	36.1
	30,001–5000	63	30.3
	50,001–7000	38	18.3
	70,001–10,000	20	9.6
	10,001 and above	12	5.7

of the respondents (46.6%) were educated at an undergraduate level. The respondents were relatively young, mainly 21–25 years old (41.8%). The top three occupations with the largest number are students (46.6%), enterprise staff (24.5%), and freelancers (11.5%).

4.3 Data Analysis

As suggested by prior researchers, partial least squares structural equation modeling (PLS-SEM) should be selected when the goal of the research is to predict key target constructs [53]. Besides, compared to CB-SEM, PLS-SEM has advantages in data analysis when the sample size is low, and the data are to some extent nonnormal [53]. PLS-SEM also renders approaches to examining the mediating effect [54]. The dataset in this study is small, and this research needs to test the given model and analyze the mediating role of patriotism, PLS-SEM thus was employed to evaluate the measurement and construal model by utilizing SmartPLS v.3.3.2.

Table 3 Reliability and validity

Items	Mean	S.D	Loading	Cronbach's Alpha	CR	AVE
FE1	6.058	1.017	0.837	0.888	0.917	0.688
FE2	5.904	1.088	0.846			
FE3	5.817	1.199	0.796			
FE4	5.870	1.113	0.874			
FE5	5.582	1.291	0.793			
PTT1	6.702	0.732	0.936	0.933	0.957	0.882
PTT2	6.750	0.632	0.945			
PTT3	6.683	0.731	0.936			
STB1	5.529	1.297	0.712	0.861	0.906	0.709
STB2	5.938	1.101	0.900			
STB3	5.894	1.164	0.860			
STB4	5.976	1.116	0.883			
RI1	6.106	1.073	0.854	0.894	0.926	0.758
RI2	6.192	0.957	0.880			
RI3	6.096	1.065	0.898			
RI4	5.870	1.224	0.850			

5 Results

5.1 Reliability and Validity

As shown in Table 3, the values of loadings, Cronbach's alpha, composite reliability (CR), and average variance extracted (AVE) for all the constructs are higher than 0.7, 0.7, 0.7, and 0.5, respectively [53]. The square root of AVE for each construct is greater than its correlations with any other construct [55]. Moreover, as shown in Table 4, the heterotrait–monotrait ratio (HTMT) ranges from 0.374 to 0.765 (lower than 0.85). The reliability and validity of the measurement model are therefore satisfactory [56].

5.2 Hypothesis Testing

Before estimating the structural relationships in the given model, the collinearity issue was examined to ensure the regression results without common method bias (CMB). Harman's single-factor test shows that the variance percentage of the first factor without rotation was 47.676%, lower than 50% (Podsakoff et al. 2012). Furthermore, the values of all VIF (ranging from 1.000 to 1.147) were lower than 3.3 [54].

Table 4 Discriminant validity

Construct	Fornell–Larcker criterion				Heterotrait–Monotrait ratio		
	FE	PTT	STB	RI	FE	PTT	STB
Flow experience (FE)	0.830						
Patriotism (PTT)	0.358	0.939			0.375		
Storytelling behavior (STB)	0.573	0.343	0.842		0.648	0.374	
Revisit intention (RI)	0.663	0.372	0.671	0.871	0.723	0.410	0.765

Therefore, the CMB is not a problem in the current study. The coefficients of determination (R^2) and cross-validated redundancy (Q^2) were further applied to check the model's explanatory power. The values of R^2 for storytelling and revisit intention were 0.350, and 0.460, respectively, indicating that the explanation levels of variance of those two endogenous constructs are "moderate" [54]. All the values of Q^2 were larger than 0.5, denoting the model has large predictive relevance [54].

The results of the PLS-SEM analysis are shown in Table 5, flow experience positively affects patriotism ($\beta = 0.358$, $p < 0.001$), storytelling behavior ($\beta = 0.516$, $p < 0.001$), and revisit intention ($\beta = 0.608$, $p < 0.001$). Patriotism positively affects storytelling behavior ($\beta = 0.158$, $p < 0.05$) and revisit intention ($\beta = 0.154$, $p < 0.01$). The f^2 effect size for significant paths ranged from 0.033 to 0.596, and the lowest value was higher than 0.02. The results indicated that the effects of exogenous constructs on endogenous constructs in the present study were acceptable, ranging from small to large [57].

Table 5 Hypothesis testing results

Path	Coefficient (β)	p	f-square	VIF	95% confidence interval		Support?
					2.50%	**97.50%**	
FE→PTT	0.358	0.000	0.147	1.000	0.221	0.490	Yes
FE→STB	0.516	0.000	0.357	1.147	0.387	0.620	Yes
FE→RI	0.608	0.000	0.596	1.147	0.489	0.701	Yes
PTT→STB	0.158	0.017	0.033	1.147	0.027	0.286	Yes
PTT→RI	0.154	0.009	0.038	1.147	0.045	0.271	Yes

FE, flow experience; PTT, patriotism; STB, storytelling behavior; RI, revisit intention

Table 6 The results of mediating effects

Path	Coefficient	p	95% confidence interval		Support?
			2.50%	97.50%	
FE → PTT → STB	0.057	0.041	0.010	0.120	Yes
FE → PTT → RI	0.055	0.036	0.014	0.118	Yes

FE, flow experience; PTT, patriotism; STB, storytelling behavior; RI, revisit intention

5.3 Mediating Effects

The results of mediate effects are shown in Table 6. The results indicate that patriotism mediates the effect of flow experience on storytelling behavior ($\beta = 0.057, p < 0.05$, CI[0.010, 0.120]), while patriotism mediates the effect of flow experience on revisit intention ($\beta = 0.055, p < 0.05$, CI[0.014, 0.118]).

6 Conclusion, Discussion, and Implications

6.1 Conclusion

This study tests a model linking flow experience, patriotism, storytelling behavior, and revisit intention. Flow experience has positive effects on patriotism, storytelling behavior, and revisit intention. Patriotism has positive effects on storytelling behavior and revisit intention. Patriotism mediates the effects of flow experience on storytelling behavior and revisit intention. Therefore, this study establishes the relational flow from flow experience to storytelling behavior and revisit intention as "flow experience → patriotism → storytelling behavior" and "flow experience → patriotism → revisit intention", respectively. It can be concluded that flow experience can trigger tourist patriotism, which in turn promotes their storytelling behavior and revisit intention. This study contributes to our understanding of tourist patriotism and their inheritance behavior of red culture.

6.2 Theoretical Implications

This study sheds light on the trace that the direction of red tourism research has shifted from a government-led/supplier-led perspective to a tourist-led perspective, encouraging scholars to note that red tourism is market-oriented at the current development stage and to pay attention to red tourists' psychology and behavior. Accordingly, this study makes a meaningful attempt to explore red tourists' cognitive and behavioral changes caused by the red tourism experience. Our research responds to previous calls by researchers to focus on investigating the psychological and behavioral mechanisms of red tourists [2].

Second, this study highlights tourists' cognition changes (manifested by patriotism) in the context of red tourism by quantitatively measuring patriotism and examining its relationship with other relative variables. Our work is a pioneer study that provides quantitative evidence to support the association between red tourism and tourist patriotism. The findings of this study, in conjunction with previous qualitative research [58, 59], help to demonstrate the whole picture of how red tourism affects tourists' cognition, emotion, and behavior towards the party/nation/nation.

Finally, we identify and illustrate tourist storytelling behavior and revisit intention in the background of red tourism based on real-world phenomena and literature. The results of PLS-SEM indicate that these two favorable post-trip behaviors (intention) are affected by flow experience and patriotism, which further prove and display tourists' active co-creator roles in experience co-creation as well as value co-creation for this industry [7]. By doing that, this study service an observer of tourists' role in the inheritance of red culture embedded in red tourism destinations.

6.3 Practical Implications

Providing a good red cultural atmosphere to facilitate flow experiences is an alternative way for stimulating tourists to construct patriotism when traveling to a red destination [4]. For creating an experiential atmosphere, a good "scene" and "script" are very important [60]. Destination management organizations can create distinguishable themes according to the development sequence of historical events and the life trajectories of historical personages, supplemented by a variety of modern performance forms, such as ppt, projection, 3d, and VR [61] and activities (i.e., jubensha, wear special clothes), to shape static and dynamic red cultural stimuli sources [4].

Additionally, Destination management organizations should hold the awareness that tourists are the optimum communicators of red culture. As such, managers ought to develop strategies to attract tourists to share their red tourism experiences online or in the real world and visit again. For instance, encouraging tourists to post pictures or videos related to their experience of red tourism on their social networking sites by giving free tickets to revisit the same red tourism destination or visit other similar red tourism destinations. This approach promotes both tourist storytelling behavior and tourist loyalty intention or behavior.

6.4 Limitations and Future Research

This study has at least two limitations that further work may be to be improved. First, in the literature, the inner construct of patriotism has not achieved an agreement. Although the current study considers patriotism as a unidimensional term following

Choe et al. [19], more scholars suggest that patriotism is a multi-dimensional concept [21], thus, researchers should take insight to understand the inner construct when conducting an empirical study in the future. Second, this study only employs flow experience as the trigger factor of patriotism; future studies, therefore, can advance patriotism research by identifying various elements that might affect patriotism in multiple research contexts.

References

1. Zahra, A., McIntosh, A.J.: Volunteer tourism: evidence of cathartic tourist experiences. Tour. Recreat. Res. **32**(1), 115–119 (2007)
2. Zhao, S.N., Timothy, D.J.: Tourists' consumption and perceptions of red heritage. Ann. Tour. Res. **63**, 97–111 (2017)
3. Pretes, M.: Tourism and nationalism. Ann. Tour. Res. **30**(1), 125–142 (2003)
4. Tang, W., Zhang, L., Yang, Y.: Can red tourism construct red memories? Evidence from tourists at Mount Jinggang, China. J. Destin. Mark. Manag. **20**, 1–11 (2021)
5. Zuo, B., Huang, S., Liu, L.: Tourism as an agent of political socialisation. Int. J. Tour. Res. **18**(2), 176–185 (2016)
6. Zhong, S., Lu, W., Peng, H., Dai, S.: Red scarf children's cognition of red tourism resources and patriotism: evidence from Yuhuatai Revolutionary Martyrs Memorial Area in Nanjing. J. Nat. Resour. **36**(7), 1684–1699 (2021)
7. Pera, R.: Empowering the new traveller: Storytelling as a co-creative behaviour in tourism. Curr. Issue Tour. **20**(4), 331–338 (2017)
8. Wang, J. W.: (2021). Annual report on the development of red tourism in China (2021). Social Science Academic Press (China)
9. Li, M., Cai, L.A., Lehto, X.Y., Huang, J.: A missing link in understanding revisit intention—The role of motivation and image. J. Travel Tour. Mark. **27**(4), 335–348 (2010)
10. Hsu, C.H., Cai, L.A., Wong, K.K.: A model of senior tourism motivations—anecdotes from Beijing and Shanghai. Tour. Manage. **28**(5), 1262–1273 (2007)
11. Liu, R., Huang, Z., Yu, R., Bao, J., Mo, Y.: The impact of red tourism on national identity of tourists. J. Nat. Resour. **36**(7), 1673–1683 (2021)
12. Song, C., Li, X., Zhang, Q.: The high-quality development of red tourism. Tourism Tribune **36**(6), 3–5 (2021)
13. Zhou, X., Wong, J.W.C., Wang, S.: Memorable tourism experiences in red tourism: the case of Jiangxi, China. Front. Psychol. **13**, 899144 (2022)
14. Zhou, M., Yan, L., Wang, F., Lin, M.: Self-congruity theory in red tourism: a study of Shaoshan city, China. J. China Tourism Res. **18**(1), 46–63 (2022)
15. Doob, L.W.: Patriotism and nationalism: their psychological foundations. Am. Behav. Sci. **8**(3), 27–27 (1964)
16. Kosterman, R., Feshbach, S.: Toward a measure of patriotic and nationalistic attitudes. Polit. Psychol. **10**(2), 257–274 (1989)
17. Sidanius, J., Feshbach, S., Levin, S., Pratto, F.: The interface between ethnic and national attachment: ethnic pluralism or ethnic dominance? Public Opinion Quart. **61**(1), 102–133 (1997)
18. Brown, G., Smith, A., Assaker, G.: Revisiting the host city: an empirical examination of sport involvement, place attachment, event satisfaction and spectator intentions at the London Olympics. Tour. Manage. **55**, 160–172 (2016)
19. Choe, Y., Kim, H., Cho, I.: Role of patriotism in explaining event attendance intention and media consumption intention: The case of Rio 2016. Curr. Issue Tour. **23**(5), 523–529 (2020)

20. Kim, H., Choe, Y., Lee, C.-K.: Differential effects of patriotism and support on post-development visit intention: The Korean DMZ Peace Park. J. Travel Tour. Mark. **36**(3), 384–401 (2019)
21. Qu, Z., Cao, X., Ge, H., Liu, Y.: How does national image affect tourists' civilized tourism behavior? The mediating role of psychological ownership. J. Hosp. Tour. Manag. **47**, 468–475 (2021)
22. Green, E.G.T., Sarrasin, O., Fasel, N., et al.: Nationalism and patriotism as predictors of immigration attitudes in Switzerland: a municipality-level analysis. Swiss Polit. Sci. Rev. **17**(4), 369–393 (2011)
23. Huddy, L., Del Ponte, A., Davies, C.: Nationalism, patriotism, and support for the European Union. Polit. Psychol. **42**(6), 995–1017 (2021)
24. Stepchenkova, S., Dai, X., Kirilenko, A.P., Su, L.: The influence of animosity, ethnocentric tendencies, and national attachment on tourists' decision-making processes during international conflicts. J. Travel Res. **59**(8), 1370–1385 (2020)
25. Csikszentmihalyi, M.: Beyond boredom and anxiety. Jossey-Bass Publishers, San Francisco, CA (1975)
26. Csikszentmihalyi, M.: Flow: The psychology of optimal experience. Harper and Row, New York, NY (1990)
27. Csikszentmihalyi, M.: Beyond boredom and anxiety: Experiencing flow in work and play, 2nd edn. Jossey-Bass, San Francisco, CA (2000)
28. da Silva deMatos, N. M., de Sa, E. S., & de Oliveira Duarte, P. A.: A review and extension of the flow experience concept. Insights and directions for Tourism research. Tourism Manage. Perspect. **38**, 1–17 (2021)
29. Huang, X.T., Wei, Z.D., Leung, X.Y.: What you feel may not be what you experience: a psychophysiological study on flow in VR travel experiences. Asia Pacific J. Tourism Res. **25**(7), 736–747 (2020)
30. Tang, J., Yan, L., Xu, J.: Tourists' experience of iconic public art in Macau. J. Tour. Cult. Chang. **19**(1), 79–96 (2021)
31. Zhou, X., Wong, J.W.C., Xie, D., Liang, R., Huang, L.: What does the audience care? The effects of travel vlog information quality on travel intention. Total Qual. Manag. Bus. Excell. **34**(15–16), 2201–2219 (2023)
32. Lunardo, R., Ponsignon, F.: Achieving immersion in the tourism experience: the role of autonomy, temporal dissociation, and reactance. J. Travel Res. **59**(7), 1151–1167 (2020)
33. Wu, C.H.J., Liang, R.D.: The relationship between white-water rafting experience formation and customer reaction: a flow theory perspective. Tour. Manage. **32**(2), 317–325 (2011)
34. Coble, T.G., Selin, S.W., Erickson, B.B.: Hiking alone: understanding fear, negotiation strategies and leisure experience. J. Leis. Res. **35**(1), 1–22 (2003)
35. Su, L., Pan, L., Wen, J., Phau, I.: Effects of tourism experiences on tourists' subjective well-being through recollection and storytelling. J. Vacat. Mark. **29**(4), 479–497 (2023)
36. Mei, X.Y., Hågensen, A.M.S., Kristiansen, H.S.: Storytelling through experiencescape: creating unique stories and extraordinary experiences in farm tourism. Tour. Hosp. Res. **20**(1), 93–104 (2020)
37. Sole, D., Wilson, D.G.: Storytelling in organizations: the power and traps of using stories to share knowledge in organizations. LILA, Harvard, Grad. School Educ. **9**(1), 1–12 (2002)
38. Keskin, H., Akgün, A.E., Zehir, C., Ayar, H.: Tales of cities: city branding through storytelling. J. Glob. Strat. Manag. **10**(1), 31–41 (2016)
39. Woodside, A.G., Sood, S., Miller, K.E.: When consumers and brands talk: storytelling theory and research in psychology and marketing. Psychol. Mark. **25**(2), 97–145 (2008)
40. Cater, C., Albayrak, T., Caber, M., Taylor, S.: Flow, satisfaction and storytelling: a causal relationship? Evidence from scuba diving in Turkey. Curr. Issue Tour. **24**(12), 1749–1767 (2021)
41. Ajzen, I., Driver, B.L.: Application of the theory of planned behavior to leisure choice. J. Leis. Res. **24**(3), 207–224 (1992)

42. Ulker-Demirel, E., Ciftci, G.: A systematic literature review of the theory of planned behavior in tourism, leisure and hospitality management research. J. Hosp. Tour. Manag. **43**, 209–219 (2020)
43. Li, F., Wen, J., Ying, T.: The influence of crisis on tourists' perceived destination image and revisit intention: an exploratory study of Chinese tourists to North Korea. J. Destin. Mark. Manag. **9**, 104–111 (2018)
44. Nguyen Viet, B., Dang, H.P., Nguyen, H.H.: Revisit intention and satisfaction: the role of destination image, perceived risk, and cultural contact. Cogent Business Manage. **7**(1), 1–20 (2020)
45. Back, R.M., Bufquin, D., Park, J.Y.: Why do they come back? The effects of winery tourists' motivations and satisfaction on the number of visits and revisit intentions. Int. J. Hosp. Tour. Adm. **22**(1), 1–25 (2021)
46. Kim, M., Thapa, B.: Perceived value and flow experience: application in a nature-based tourism context. J. Destin. Mark. Manag. **8**, 373–384 (2018)
47. Rutherdale, R.: Canada's August festival: communitas, liminality, and social memory. Canad. Histor. Rev. **77**(2), 221–249 (1996)
48. Ding, H.M., Hung, K.P.: The antecedents of visitors' flow experience and its influence on memory and behavioral intentions in the music festival context. J. Destin. Mark. Manag. **19**, 1–11 (2021)
49. Wong, J.W.C., Lai, I.K.W., Tao, Z.: Sharing memorable tourism experiences on mobile social media and how it influences further travel decisions. Curr. Issue Tour. **23**(14), 1773–1787 (2020)
50. Meng, B., Chua, B.-L., Ryu, H.B., Han, H.: Volunteer tourism (VT) traveler behavior: Merging norm activation model and theory of planned behavior. J. Sustain. Tour. **28**(12), 1947–1969 (2020)
51. Brislin, R.W.: Back-translation for cross-cultural research. J. Cross Cult. Psychol. **1**(3), 185–216 (1970)
52. Li, F. (2019). Understanding Chinese tourists' motivations of sharing travel photos in WeChat. Tourism Manage. Perspect. 1–13
53. Hair, J.F., Ringle, C.M., Sarstedt, M.: PLS-SEM: indeed a silver bullet. J. Market. Theory Practice **19**(2), 139–152 (2011)
54. Hair, J.F., Jr., Sarstedt, M., Hopkins, L., Kuppelwieser, V.G.: Partial least squares structural equation modeling (PLS-SEM): An emerging tool in business research. Eur. Bus. Rev. **26**(2), 106–121 (2014)
55. Fornell, C., Larcker, D.F.: Evaluating structural equation models with unobservable variables and measurement error. J. Mark. Res. **18**(1), 39–50 (1981)
56. Henseler, J., Ringle, C.M., Sarstedt, M.: A new criterion for assessing discriminant validity in variance-based structural equation modeling. J. Acad. Mark. Sci. **43**(1), 115–135 (2015)
57. Cohen.: Statistical power analysis for the behavioral sciences, 2nd edn. Lawrence Erlbaum Associates, NJ, USA (1988)
58. Liu, H., Yue, N., Bai, C.H.: The influence of emotional arousal on tourists' cognition in the context of red tourism. Social Scientist **3**, 84–90 (2018)
59. Zhao, Z.F., Sun, G.D., Li, Z.W.: Social effects of red tourism: a perspective of identity. J. Sichuan Normal Univ. (Soc. Sci. Ed.) **43**(1), 63–71 (2016)
60. Lai, I.K.W., Liu, Y., Lu, D.: The effects of tourists' destination culinary experience on electronic word-of-mouth generation intention: the experience economy theory. Asia Pacific J. Tourism Res. **26**(3), 231–244 (2021)
61. Wu, X., Lai, I.K.W.: The use of 360-degree virtual tours to promote mountain walking tourism: stimulus–organism–response model. Informat. Technol. Tourism **24**(1), 85–107 (2022)

Spatial Distribution and Utilization of Red Tourism Resource in Guangdong Province, China

Yinghui Chen, Yi Guo, and Shengchao Li

Abstract Taking the 480 red tourism resources in Guangdong Province as the research object, using the method of combining spatial analysis and SWOT analysis, we discuss the spatial distribution, influencing factors and development and utilization status of red tourism resources in Guangdong Province. Our research found that the spatial distribution of red tourism resources in Guangdong Province is in the shape of "one focus and multiple points". The overall spatial distribution is negatively correlated, and the distribution difference is significant. The overall spatial correlation degree is low while the local spatial correlation degree is high. The spatial distribution characteristics of red tourism resources in Guangdong Province are mainly affected by historical, economic and policy factors. The current situation of the development and utilization of red tourism resources in Guangdong Province is the coexistence of opportunities from the tourism market and policies and challenges from market competition. Finally, we put forward some suggestions for the development and utilization of red tourism resources in Guangdong Province.

Keywords Red Tourism Resources · Spatial distribution · Influencing factors · Development and utilization · Guangdong Province

Y. Chen · S. Li (✉)
School of Management, Guangdong University of Education, Guangzhou, Guangdong, People's Republic of China
e-mail: lishengchao@qq.com

Y. Chen
e-mail: 1274114492@qq.com

Y. Guo
School of Tourism and Business, Guangzhou Panyu Polytechnic, Guangzhou, Guangdong, People's Republic of China
e-mail: guoyi1989@foxmail.com

Y. Luo et al. (eds.), *Tourist Experience and Perception in the Pandemic and Post-Pandemic Era*, https://doi.org/10.1007/978-981-96-1820-0_13

1 Introduction

Red tourism plays an important role in China's tourism industry and has a very broad development prospect. At the end of 2004, the Chinese government released the outline of the first phase of the national red tourism development plan. Officially put forward the concept of "red tourism", that is, tourism with resources related to the history of the Communist Party of China and revolutionary sites. According to the statistics of the Ministry of Culture and Tourism of China, the number of tourists in China's red tourism has increased steadily year by year during the "13th Five-Year Plan" period. The domestic tourism market share of red tourism exceeds 11%. In 2020, there will be more than 100 million red tourism tourists nationwide. 2021 marks the centenary of the founding of the Communist Party of China, and a wave of red tourism fever has emerged in China, with red tourism activities becoming increasingly popular. The upsurge of red tourism in Guangdong has also been heating up in recent years. The abundant and high-quality red tourism resources are the important material foundation and strong support for the development of red tourism in Guangdong Province. Red tourism resource is an important part of red tourism, and the current research mainly focuses on the development and utilization of tourism resources [1], the development evaluation of tourism resources [2], and the spatial structure of tourism resources [3]. The research on red tourism resources in Guangdong Province has also received more and more extensive attention, which has greatly promoted the utilization of red tourism resources in Guangdong Province and the development of red tourism, and provided resource data and analysis methods for follow-up related research. Therefore, it is necessary to further study the spatial distribution characteristics and development and utilization of red tourism resources for red tourism in Guangdong Province.

Red tourism resources cover the location, items and spirit of the Communist Party of China. It is a product of important commemorative significance during the Revolution. Most of the resources are distributed in the Red Army Long March and the Chinese Revolution bases. Red tourist attractions are widely distributed in China and have obvious regional characteristics. [3].

Reviewing the research on China's red tourism resources, it is found that GIS and other technical means are mostly used to research red tourism resources in different regions [4], and to deeply analyze its spatial pattern and influencing factors. Zou et al. (2021) pointed out that the spatial correlation of classic red tourism attractions in China varies greatly, and the spatial distribution pattern is affected by society, environment, resources and economy [5]. Wang and Huang (2021) pointed out that the Long March red tourist attractions are clustered and distributed as a whole, showing the spatial distribution characteristics of "large concentration and small dispersion" [6]. Xin et al. (2021) pointed out that China's patriotic education bases have significant spatial concentration, diffusion and directionality [7]. Lu and Bai (2021) used GIS visualization analysis to draw the conclusion that the red tourism resources in Gansu Province have the characteristics of overall centralized distribution and uneven regional distribution in space. Tang et al. (2019) found that the spatial pattern of red tourism resources in Hunan Province is mainly affected by five factors: historical events, terrain, policies, red cultural heritage, and regional economy [8]. The above research results show that the influencing factors of the spatial distribution of red tourism resources have similarities and commonalities.

The research on the development and utilization of red tourism resources mostly uses the method of situation analysis to discuss the case. Mu Hong (2021) used the SWOT model to analyze and study the development and utilization of red tourism resources in Shidingshan, Hejiang, and proposed development strategies focusing on rural revitalization [9]. Lu Jiaying (2020) used the SWOT method to analyze the red tourism resources of Xuefeng Mountain, and suggested to enhance its competitiveness through innovative integration, strengthening of talents and facilities construction and other schemes [10]. Xiao Quanliang (2019) used the SWOT method to study the development and utilization of red cultural resources in Qinghai Province, and proposed strategies to focus on protection, highlighting advantages, functional attributes and economic value [11]. Gao Junbin and Li Jingyi (2019) conducted a SWOT analysis on the development of red tourism resources in Lvliang, and proposed a strategy for cross-regional cooperation and sharing, and the development of global tourism centered on red tourism [12]. Li Xiaomei (2018) conducted a SWOT analysis on the red tourism resources of the Hexi Corridor, and proposed that it should be placed in the background of national red tourism and provincial tourism development to expand the space for subsequent development [13]. Chen Hui (2015) conducted a SWOT analysis on the development of red tourism resources in the Dabie Mountains of Anhui from the perspective of Yuexi region, combining modern tourism and red culture to promote the sustainable development of red tourism [14]. The existing literature mainly puts forward countermeasures and suggestions for the status quo of development and utilization, and lacks consideration of the characteristics of the spatial distribution of resources.

"Red" as a political concept originated abroad, while "red tourism" is a unique innovation in China [15, 16]. The research on China's red tourism resources is a research topic with Chinese characteristics. Yang and Huang (2018) pointed out in their research that foreign scholars have rarely done relevant research on China's red tourism. However, with the vigorous development of red tourism and the purpose of in-depth understanding of Chinese red culture, foreign scholars have begun to study related fields of Chinese red tourism in recent years [17, 18].

To sum up, there are abundant literatures related to research on China's red tourism resources, and different research objects and regions have different research results. However, there is a shortage of research using the combination of spatial analysis and SWOT analysis. In order to make up for this deficiency, this study adopts a new method combining the two to study the spatial distribution and influencing factors of red tourism resources in Guangdong Province, as well as the development and utilization of red tourism resources, and put forward corresponding countermeasures and suggestions.

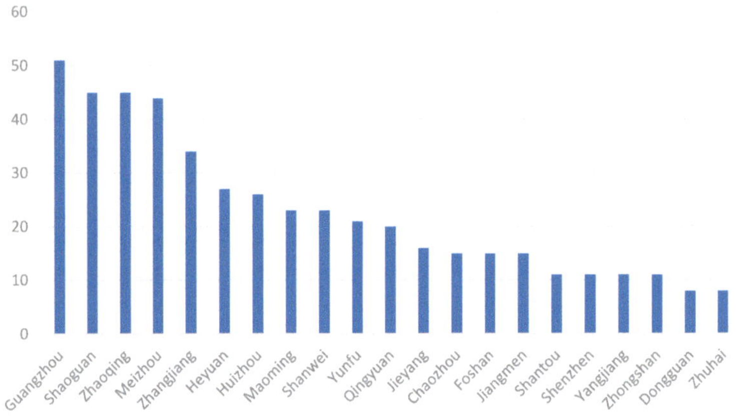

Fig. 1 The number of Red Tourism Resources in each city in Guangdong Province

2 Data collection and research methods

2.1 Data collection

The data sources of this study are related literature and network data. A total of 480 red tourism resource points in 21 prefecture-level cities in Guangdong Province were collected (Figures 1 and 2), of which 353 were from the red cultural map of Guangdong Province issued by the Department of Natural Resources of Guangdong Province. There are 11 other sources from relevant literature [19, 20], 81 sources from Boya Travel Sharing Network and China Red Tourism Network, and 35 sources from the official website of Guangdong Red Culture Research Association.

These resources are mainly composed of important red revolutionary sites and important red military cultural heritage, including sites, former residences, cemeteries, steles, parks, exhibition halls, academies, and patriotic education bases. See Table 1 for a list of resources.

The list is too long and limited in space, only the data of the first three cities are listed.

2.2 Research methods

2.2.1 Kernel density analysis

Kernel Density Estimation (KDE) is a non-parametric estimation method, which is widely used as a method for calculating the probability of a point set geographic event occurring in a target area [21]. Through the kernel density analysis tool in ArcGIS software, the density of point features or the density of line features in its

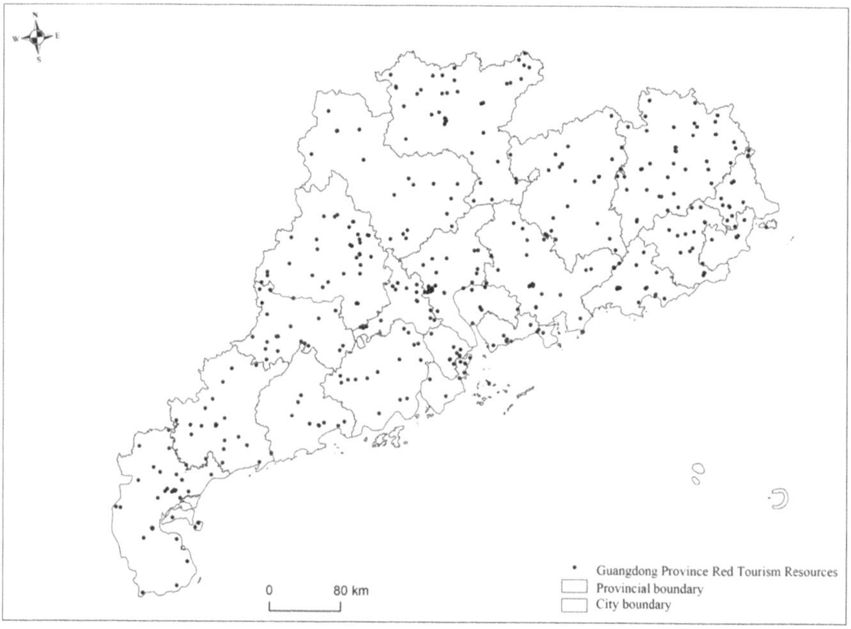

Fig. 2 Distribution map of the spatial pattern of Red Tourism Resources in Guangdong Province

surrounding area can be obtained. The density of the research objects in the research area determines the probability of its occurrence. The denser the points, the higher the frequency of events, and vice versa. The formula for this method is:

$$f_n(x) = \frac{1}{nh}\sum_{i=1}^{n} k\left(\frac{x - x_i}{h}\right) \tag{1}$$

2.2.2 Spatial Autocorrelation Analysis

Spatial autocorrelation refers to the possible interdependence between observed data of some variables in the same distribution area [22]. Moran's I is the most widely used and can be divided into global Moran index and local Moran index. The global Moran index is a variance-normalized rational number between −1 and +1. When the global Moran index is greater than 0, it indicates that the data is spatially positively correlated, and with the increase of the value, the spatial correlation is stronger; when it is less than 0, it indicates that the data is spatially negatively correlated, and it behaves as the value decreases. A larger spatial difference; when it is 0, it indicates that the spatial correlation of the data is random. When parsing the global Moran index, you need to combine the P-value and Z-score. The local Moran index

Table 1 List of 480 Red Tourism Resources in Guangdong Province

City	Number	Details of red tourism resources
Qingyuan	20	Qinhuangshan Base Monument, Fogang County Revolutionary Martyrs Cemetery, Yingde City Martyrs Cemetery, Lianzhou Revolutionary Martyrs Cemetery, Yangshan Revolutionary Martyrs Cemetery, Yatang Primary School, Feng Dafei's Former Residence, Chongziyan, Chenghuang Street Guanyin Hall, Yingyangguan Revolution Ruins, Yixian Deng Gong Temple, Qingcheng District Siyuan Garden, Datong Club, the former site of the Guangdong-Guangxi-Xiangbian Column Headquarters of the Chinese People's Liberation Army, the former site of the Eastern Longitudinal Beijiang Detachment Headquarters, Mo's Mansion, Qingyuan County Revolutionary Martyrs Monument, and the former site of Hui'ai Hospital, Xiangxiu Memorial Park, Xilian County Anti-Japanese War Memorial Hall
Dongguan	8	Dalingshan Revolutionary Martyrs Monument, the former site of Guangdong People's Anti-Japanese Guerrilla Dongjiang Column Road East Cadre Training Course, Xiong's Ancestral Hall, Guangdong Dongjiang Column Memorial Hall, "Seven Warriors of Love for the People and Rescue" Monument, Dalingshan Anti-Japanese Base Site Site, Dongguan Baihuadong Battle Site, Historic Site of Shilong Zhongshan Park in Dongguan City
Chaozhou	15	Chaochengrao Revolution "One Hometown", Hanbi Building, Koji Temple, Phoenix Mountain Revolutionary Memorial Park, Nanchang Uprising Army Maozhi Meeting Site, Li Liying's Former Residence, Chao'an County Xilin Township Farmers Association Site, Chaoyuan Building, Chaozhou Revolutionary Martyrs Monument, Huang Qiufu Martyr's Monument, Zhu De Maozhi Military Conference Site, Xianzhou Huang's Ancestral Hall, Guiban Woyun Thatched Cottage, Gangke Gong Temple, Yongan Village

reflects the spatial agglomeration degree of similar attribute values around an area, and shows the relationship between a specific area and neighboring areas through a LISA map. The local Moran index can also be used as the regional components of the global Moran index, which can reflect the influence of spatial autocorrelation in different regions on the spatial autocorrelation of the entire study area.

2.2.3 SWOT analysis

SWOT analysis is a systematic analysis method in management. Based on the status quo of the spatial distribution of red resources in Guangdong Province, this paper analyzes and systematically evaluates the status quo of its spatial development and utilization from four aspects: internal advantages, disadvantages, and external opportunities and threats. Finally, according to the actual situation, corresponding effective suggestions and countermeasures are put forward.

3 Empirical Analysis

3.1 Analysis of the spatial distribution characteristics of red tourism resources in Guangdong Province

Huang and Song (2013) mentioned in their research that the material elements of red tourism resources are composed of revolutionary historical sites or sites and revolutionary memorial venues [23]. Fang and Deng (2013) divided the basic types of red tourism resources into five types: hometown of great men, sites of revolutionary historical events and activities, tombstones (groups) and martyrs' cemetery, comprehensive revolutionary history memorial hall, oral and intangible cultural heritage of revolutionary history [24]. In this paper, 480 important red tourism resources in Guangdong Province are collected as the research object with material elements as the most important component and the above contents as the basic type, and the analysis is carried out from the following three parts.

3.1.1 "One-Center, Many-Points" Clustered Distribution

As shown in Figure 3, from the perspective of the four major regions, the spatial distribution density of red tourism resources in the Pearl River Delta is the highest. The red tourism resources in Guangdong Province are rich in Guangzhou and the surrounding area, and the layout is also the most concentrated. And from Guangzhou as the central area, the distribution pattern of "one center and multiple points" is presented to the northeastern region of Zhaoqing in the Pearl River Delta, the central region of Shaoguan and the western region of Yunfu in northern Guangdong, the prefecture-level administrative center of Zhanjiang in western Guangdong, and the Chaoshan region in eastern Guangdong. There are obvious differences in the overall spatial distribution density.

Within the scope of the province, the spatial distribution density of red tourism resources in Guangzhou is the highest, with a density value of 103.03–145.96 pieces/10,000 km^2. Followed by Foshan City. Again, Zhongshan City, Huizhou City, Yunfu City, Shaoguan City, Zhanjiang City and Chaoshan District. The lowest value is distributed in Qingyuan City, which is not higher than 8.01 pieces/10,000 km^2. The general trend is to disperse from the high-density areas to the surrounding low-density areas. At the same time, it also shows that the spatial distribution density of red tourism resources in each city in Guangdong Province has a strong correlation with the important revolutionary historical events in each city. In the places where important revolutionary historical events occur, there will be obvious aggregation and distribution of red tourism resources. The most typical example is Guangzhou, the birthplace of the revolution, which has the largest number of red tourism resources and the highest distribution kernel density value.

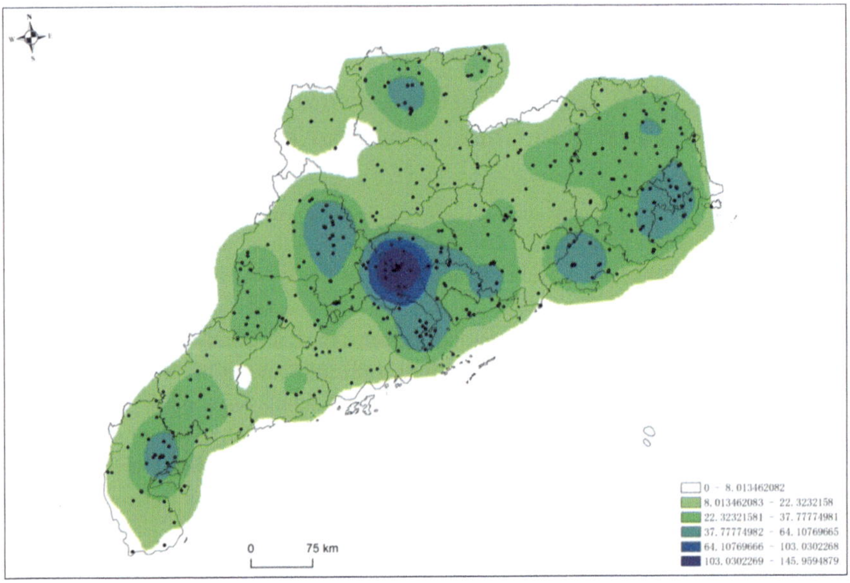

Fig. 3 Kernel density map of spatial distribution of Red Tourism Resources in Guangdong Province

3.1.2 The Spatial Distribution is Negatively Correlated, and Varies Greatly

Using the spatial autocorrelation tool in GeoDa software, the spatial correlation analysis of 480 red tourism resources in Guangdong Province was carried out, and the Moran index of the spatial distribution of red tourism resources in Guangdong Province was calculated. As shown in Fig. 4, the abscissa axis is the observed value of the space unit itself, the ordinate axis is the average value of the observed values of the adjacent units in the space, and the slope of the oblique line is the global Moran index value.

The value of the global Moran index of the spatial distribution of red tourism resources in Guangdong Province is –0.323. This indicates that the distribution of red tourism resources in Guangdong Province is negatively correlated in space. With the dispersion of the spatial distribution of red tourism resources in Guangdong Province, the correlation will become significant. At the same time, it also shows that the distribution of red tourism resources in Guangdong Province is quite different in space. The distribution of red tourism resources is not highly correlated with the number of resources in each city. This shows that Guangdong Province has time-series consistency in the process of red tourism resource development and regional development. In general, the economically developed areas of Guangdong Province have a greater impact on the diffusion effect. And the red tourism resources in this area can also be fully and fully developed and utilized.

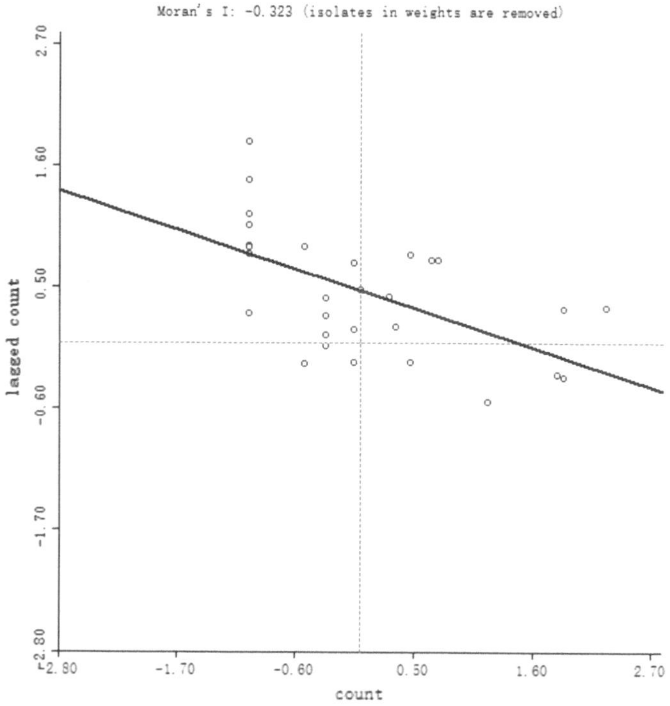

Fig. 4 The global Moran index of the spatial distribution of Red Tourism Resources in Guangdong Province

3.1.3 The Overall Layout Spatial Correlation Degree is Low, and the Local Spatial Correlation Degree is High

Using the spatial autocorrelation tool in GeoDa software to analyze the local spatial autocorrelation of 480 red tourism resources in Guangdong Province, the LISA spatial cluster map of the spatial distribution of red tourism resources in Guangdong Province was obtained. As shown in Figure 5, the spatial correlation types of tourism resources in Guangdong Province are divided into double-high area (H–H), double-low area (L–L), low-high area (L–H) and high-low area (H–L).

The H–H area refers to the area with high density of its own red tourism resources and the surrounding area. The red tourism resources in these areas have strong spatial correlation and are mainly composed of Jieyang City, Shanwei City, Huizhou City and Heyuan City. The L–H area refers to the area where the density of red tourism resources is low and the density of classic scenic spots in the surrounding area is high. It consists of Foshan City. No city is classified as L–L or H–L, and the rest are non-significant areas. On the whole, the spatial correlation of red tourism resource density values in Guangdong Province is different. The distribution range of the H–H area is wider, the distribution range of the L–H area is smaller, and the distribution range of the non-significant area is the most extensive. This shows that in a certain

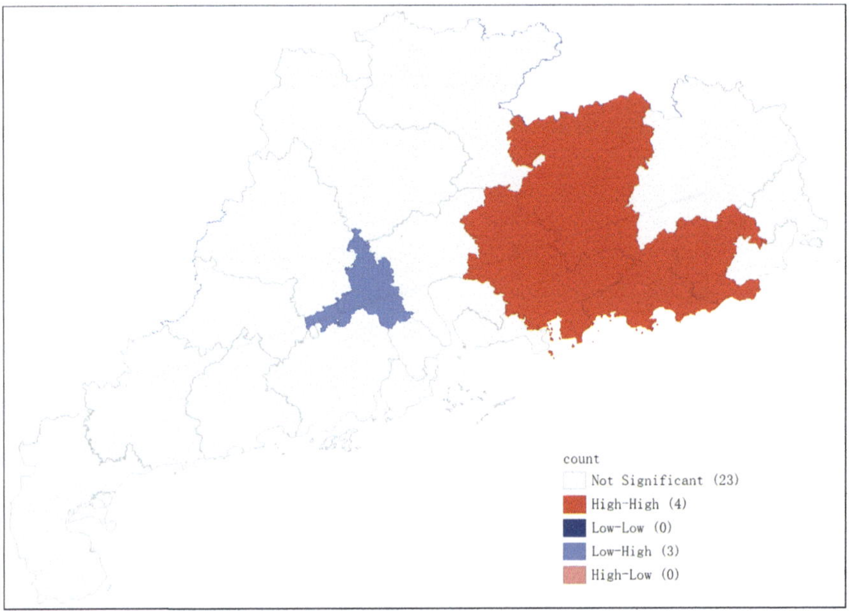

Fig. 5 LISA spatial clustering map of the spatial distribution of red tourism resources in Guangdong Province

area, giving full play to the driving effect of prominent districts and carrying out coordinated development between cities is an important way for red tourism planning and construction.

3.2 Analysis of influencing factors on the spatial distribution of red tourism resources in Guangdong Province

3.2.1 Major Revolutionary Historical Events

Red tourism resources were formed in the revolutionary era. China's revolutionary holy sites are closely related to major events in the process of China's revolution [25]. It can be seen that the spatial distribution of red tourism resources in various places is affected by major revolutionary historical events. The spatial distribution of red tourism resources in Guangdong Province is no exception. Guangdong Province is one of the important sources for the Communist Party of China to carry out its revolutionary cause, and it is also one of the key activity areas in the country. It has experienced many major revolutionary events, carried revolutionary forces and promoted the development of the Chinese revolution. From the period of the Revolutionary War to the period of development and construction, there are countless historical events in Guangdong Province, big and small. The occurrence of these revolutionary historical events will bring valuable heritage and other wealth to the local area. There are 4,189 sites of major historical events in Guangdong Province, which constitute many famous revolutionary cities. Guangzhou is the most representative revolutionary city. Various types of sites, including revolutionary sites, memorial halls, tombs of martyrs, former residences, etc., together constitute the red tourism resources of Guangdong Province.

3.2.2 Regional Economic Development Level

The level of regional economic development has an important impact on the industrialization development process of tourism. While the regional economy is growing, the living standards of the general public are rising steadily, and consumer demand is becoming more and more vigorous, developing in the direction of diversification and individualization. The increasingly perfect infrastructure and the increasingly developed transportation system are conducive to the industrialized development of tourism. In recent years, China's red tourism has continued to develop and progress, and has achieved significant political, social and economic benefits. The development of red tourism can promote the protection of red tourism resources and the continuation of red cultural genes. Its development is also inseparable from the support of regional economic forces. The more prosperous areas of red tourism development in Guangdong Province are mainly concentrated in the Pearl River Delta region. The level of economic development in the region is thriving, with well-connected transportation and complete infrastructure. Therefore, it is convenient to organize more red tourism activities. With the active participation of local residents, the regional government will pay more attention to the development and protection of red tourism resources, resulting in more distribution of resources. Although the western and northern Guangdong regions have abundant and high-quality red tourism resources. However, the economic development level of the region is different from that of the Pearl River Delta region. The transportation and other infrastructure in some cities and towns are not yet complete, making it difficult for them to independently carry out red tourism activities. As a result, the red tourism resources in some areas cannot be well protected, developed and utilized, and the distribution of resources is relatively small.

3.2.3 Protection and Development of Relics

Relevant policies on the protection and development of tourism resources have an important impact on the distribution of tourism resources. There are as many as 4,189 revolutionary sites in the province, and there are 450, 214, 52 and 15 cultural relics protection units at the county, city, provincial and national levels respectively, accounting for a total of 17.46%. There are 3458 undecided, accounting for 82.54%. Among them, Guangzhou has a total of 7 national cultural relics protection units, which is the city with the largest number. By evaluating the preservation status of 4,189 revolutionary sites, 588 are in good condition, accounting for 14.04%. The study found that the revolutionary sites that are listed as cultural relics protection units have relatively impressive preservation conditions. On the contrary, most of its preservation conditions are not optimistic, some have been damaged to varying degrees or have been used for other purposes, and some may have disappeared. Due to the different degrees of protection given to revolutionary sites by cities and differences in the actual utilization rate of revolutionary sites, the spatial distribution of red tourism resources in Guangdong Province is uneven. In recent years, the development and utilization of red tourism resources in Guangdong Province has been attached great importance to by local party committees and governments, and various channels have been used to raise funds to develop red tourism resources. The development and protection of red tourism resources have made a gratifying start. The census found that there are a large number of revolutionary sites in the Pearl River Delta region that are well-protected, the attitude towards developing red tourism is positive, and the actual utilization rate of the revolutionary sites is high. While Zhanjiang and Shaoguan have a large number of revolutionary sites, they also have the largest number of revolutionary sites with poor protection. The enthusiasm for red activities is not strong, and the utilization of revolutionary sites is not ideal.

3.3 SWOT Analysis of the Development and Utilization of Red Tourism Resources in Guangdong Province

3.3.1 Advantages

Guangdong Province is rich in red tourism resources, and the red culture is deeply rooted in the hearts of the people. From the era of the Revolutionary War to the era of modernization, red culture has always run through Guangdong, and Guangdong Province is one of the important participants. Therefore, Guangdong's red culture has many contents, rich connotations, and a wide range of distribution, forming quite rich red cultural resources. According to the survey, there are about 4,300 revolutionary sites in Guangdong Province (including 313 red military cultural heritage sites) [26], and a large number of literary works about red culture are presented and circulated in various forms. It is such a strong red cultural heritage that gives birth to precious red tourism resources, which are numerous, widely distributed and of all types. In terms of tourist attractions, Guangdong Province has 13 national red tourism classic scenic spots and 10 national 4A-level or above red tourist attractions. In terms of patriotism education bases, there are 13 and 104 at the national and provincial levels respectively. In terms of national defense education demonstration bases, there are 9 and 16 at the national and provincial levels respectively. In other respects, there are 16 provincial red tourism demonstration bases, 2 national Hong Kong and Macao youth study tour bases, and countless other red memorial halls and museums. Guangdong

Province used to be the forefront of the Chinese revolution. Rich and high-quality red tourism resources are widely distributed in the province. Each city has its own unique red tourism resources in different quantities.

Guangdong Province has an advantageous geographical location, located in a low latitude zone, with both land and sea. The north is dominated by mountainous terrain, the south is dominated by plain terrain, and the east and west are mostly hilly. The water system is dominated by the Pearl River and is rich in water resources. Most areas belong to the subtropical monsoon climate, and a small part belongs to the tropical monsoon climate. Summers are hot and hot with plenty of rain, and winters are warm with less precipitation. Appropriate natural environment and climatic conditions are conducive to the development of red tourism resources. At the same time, Guangdong Province is the southern gate of the motherland, backed by a vast inland hinterland, facing the South China Sea, and adjacent to Hong Kong and Macao. Its "window" function is not only conducive to the exchange of domestic and foreign tourism development experience, but also conducive to attracting outstanding talents in the field of tourism resource development and utilization.

3.3.2 Disadvantages

The overall grade of red tourism products is low. Among such rich red tourism resources, there is only one national 5A red scenic spot in Guangdong Province, that is, Sun Yat-sen's Hometown Tourist Area in Zhongshan. No scenic spot in Guangdong Province has been included in the list of 12 key red tourist attractions in the country, only the red tourist areas of Hunan, Jiangxi and Fujian and the red tourist areas of the Zuojiang River radiate to the Pearl River Delta region. The route of Guangdong Province was not selected as the first batch of 30 national red tourism boutique routes proposed in the first phase of the national red tourism development plan. In the first batch of national red tourism classic scenic spots, Guangdong Province initially had only 3 [27], but it has only increased to 13 in recent years. There are a total of 544 national patriotic education demonstration bases in the seven batches, of which only 13 are in Guangdong Province, and the proportion is obviously low. Although Guangdong Province has rich and unique red tourism resources with high development value, the development and utilization effect is not particularly ideal. Most of them have not become nationally renowned brands, and the overall grade of red tourism products is low. The degree of industrialization of red tourism is not high, the economic benefits are not obvious [27], and it is more difficult to form a trend of agglomeration development.

Red tourism in Guangdong Province lacks brand advantages. The abundant red tourism resources advantages of Guangdong Province have not been transformed into their own brand advantages. No matter where the tourism product is, it needs a tourism brand that highlights its own advantages, and the same is true for red tourism. A tourism brand is a combination of tourism products, services, images, names and their logos, which reflects the uniqueness of tourism products and the sense of identity of tourists. After years of development and development, many places have created well-recognized red tourism brands. For example, Guizhou has a tourism brand of "Long March Culture", Shanghai has a tourism brand of "One Congress Site", Jiangxi has a tourism brand of "Red Cradle", and Shaanxi has a tourism brand of "Revolutionary Holy Land". The red tourist attractions in Guangdong Province are scattered, but a representative, comprehensive and influential red tourism brand has not yet been created. Only sporadic scenic spots were selected into the top 50 brands of red cultural scenic spots in 2020. In the 2021 list of "100 Red Tourism Routes for the Founding of the Party", only 3 routes from Guangdong Province are included. Guangdong Province lacks the advantages of red tourism brands. At present, it is still difficult to create a tourism brand with market recognition.

3.3.3 Opportunities

The red tourism market continues to heat up. With the rise of red tourism, the popularity of red tourism across the country is still unabated. On the 70th birthday of the People's Republic of China and the centennial of the founding of the Communist Party, the red tourism market in Guangdong Province continued to heat up on the 20th National Congress of the Communist Party of China. According to the information of the Ministry of Culture and Tourism and related news, the International Labor Day holiday, the National Day holiday, and the anniversary of the founding of the Party, the Army or important historical events are all peak seasons for red tourism activities, and the number of tourists is also increasing. In addition, during the Qingming holiday in 2021, the province's classic red tourism scenic spots received a total of 488,000 tourists. During this period, the memory of the martyrs has become a common practice, and the way of memory has gradually diversified. More and more citizens and tourists go to the Cemetery of Revolutionary Martyrs, Memorial Halls of Revolutionary Events, Old Sites of Revolutionary Buildings and Red Rural Tourist Spots on different holidays to carry out corresponding festival activities.

Strong support from national policies. As the people's demand for red culture continues to increase, the party and the state have paid more and more attention to the dissemination of red culture and the protection and utilization of red revolutionary sites, and have increased policy support for the development of red tourism. Various patriotic education activities have promoted the development of red tourism and red tourism resources to a certain extent. The state promulgated the "National Red Tourism Development Plan Outline". The government played a central role in the outline, guiding the work towards the goal of vigorously developing red tourism. Immediately afterwards, relevant departments conducted discussions on the development and utilization of resources, and formulated the "National Red Tourism Classic Scenic Spot List". Among them, successful and effective measures can be used for reference in scenic spots around the world. The rapidly developing red tourism has once again become a bigger hot spot and focus in tourism consumption. This greatly promotes the development and utilization of red tourism resources. In addition, the Guangdong provincial government has also actively carried out the project of protection and utilization of revolutionary cultural relics and strengthened the construction of red tourism infrastructure. Invest more support in the red tourism boutique scenic spot system and routes, and promote the combination of red tourism poverty alleviation and rural revitalization. Make full use of red art works for publicity and promotion. Various measures taken by the Guangdong Provincial Government are all activating the utilization of Guangdong's red resources. Promote the healthy development of the industry and better protect, develop and utilize local red tourism resources.

3.3.4 Challenges

Homogeneous market and product competition is fierce. The fierce competition of homogeneous markets and products in other provinces is the biggest challenge this province has to deal with in the development and utilization of red tourism resources. Although Guangdong Province is one of the cradles of China's modern revolution and has a long history of red revolution, it is slightly inferior to other important cradles of Chinese revolution such as Beijing, Shanghai, Jiangxi, Shaanxi, Anhui and so on. Most of the provinces and regions around Guangdong have formed a certain scale and have high-profile red tourism brands, seizing the opportunity of the red tourism market. This has created fierce competition with Guangdong Province, which does not have a mature and influential red tourism brand. It has caused a certain pressure on the development and utilization of red tourism resources in the province and created a huge challenge. The province is facing the challenge by creating a red tourism brand of "Million Miles of Red Road".

To sum up, the development and utilization of red tourism resources in Guangdong Province is generally good, with outstanding resource advantages. At the time of opportunity in the tourism market and national policies, Guangdong Province is constantly overcoming the disadvantages of tourism products and brands in meeting challenges.

4 Conclusions and recommendations

4.1 Conclusions

With the help of spatial analysis and SWOT analysis methods, this paper studies its spatial distribution, influencing factors and development and utilization status based on the data of red tourism resources in Guangdong Province. The main research conclusions are as follows:

(1) The overall spatial distribution of red tourism resources in Guangdong Province is in the shape of "one center and multiple points". The region with the highest distribution density is the Pearl River Delta region, and the spatial distribution density of red tourism resources in Guangzhou is the highest. Followed by

Foshan City, Zhongshan City, Huizhou City, Yunfu City, Shaoguan City, Zhanjiang City and Chaoshan District. The lowest value is distributed in Qingyuan City, and it shows a trend of spreading from the high-density areas to the surrounding low-density areas. In addition, its distribution presents a negative spatial correlation, and the regional distribution has obvious differences in space. The overall layout has a low degree of spatial correlation, and a high degree of local spatial correlation, with obvious spatial aggregation. Red tourism resources are clustered and distributed along the places where important revolutionary historical events occurred, but the degree of correlation with the number of resources in each city is not high.

(2) Major revolutionary historical events are the most obvious and direct influencing factors on the spatial distribution of red tourism resources in Guangdong Province. While promoting and promoting the development of red tourism resources in Guangdong Province, policies related to the protection and development of relics also affect their spatial distribution to a certain extent. Regional economic conditions also play a role. In the long course of revolutionary struggle, the Communist Party of China has already led and carried out many important revolutionary activities and struggles in Guangdong Province. Many famous leaders and revolutionaries have been active in Guangdong. The influence of these events has left a lot of red tourism resources for the local area. The improvement of the regional economic development level and the relevant local policies on the protection and development of relics can make full use of the local red tourism resources and make them better preserved. It is the joint influence of various aspects that has created the current spatial layout of red tourism resources in Guangdong Province.

(3) The status quo of the development and utilization of red tourism resources in Guangdong is a situation in which advantages and disadvantages, opportunities and challenges coexist. It has the advantages of rich tourism resources and advantageous geographical location. There are disadvantages such as low overall grade of red tourism products and lack of brand advantages of red tourism. Facing the opportunity of the continuous heating up of the red tourism market and the strong support of national policies. It needs to deal with the challenges of fierce competition in the homogeneous market in other provinces. The red tourism in Guangdong Province has good development conditions and foundation. At present, the development and utilization of red tourism resources and the development of red tourism show a considerable trend. The development and utilization of red tourism resources will be continuously optimized. Even though the development of red tourism faces certain difficulties, it still has huge potential and broad room for improvement.

4.2 Policy Recommendations

According to the above research conclusions, in view of the spatial distribution and development and utilization of red tourism resources in Guangdong Province, this paper adopts the SO strategy. Relying on internal advantages and taking advantage of external opportunities, the countermeasures and suggestions for further personalized development and full utilization of red tourism resources in Guangdong Province are put forward:

(1) Maintain the advantages of red tourism resources in Guangdong Province and give full play to the geographical advantages. Strengthen the protection of revolutionary cultural relics and revolutionary relics by carrying out the protection and utilization project of revolutionary cultural relics, incorporating the protection of revolutionary relics into urban planning and construction, and repairing damaged revolutionary relics, to maintain the quantitative advantage of red tourism resources. At the same time, the red tourism resources will be enriched and improved by integrating the construction of memorial parks, erecting monuments for revolutionary sites, and strengthening the construction of memorial facilities and venues. Give full play to the superiority of the geographical location of Guangdong Province, actively respond to the national issue of red tourism, and realize the optimal combination of red tourism resources in a more practical way.
(2) Improve the overall grade of red tourism products and create a red tourism brand with distinctive characteristics. Guangdong's red tourism products and brands need the combined strength of all aspects. New elements should be added to high-quality routes, the construction of key engineering projects should be implemented, and the construction of high-quality scenic spots should be accelerated. Build red research bases and patriotic education positions at various sites. In this way, we will optimize the construction of the province's red tourist attractions system, and create red tourism themed sections by integrating the resources of the area, such as Guangfo, Zhongjiang Fozhu, Guanshen Huizhao, Shanhehui, Chaoshan Jiemei, Shaoqing, etc. Adjust and optimize the construction of red tourist areas in a way that highlights the center of gravity. For example, the construction of the eastern area is centered on Chaozhou and Meizhou. The construction of the central area is centered on Guangzhou and Shenzhen, the construction of the western area is centered on Zhanjiang, and the construction of the northern area is centered on Shaoguan. At the same time, make full use of carefully developed tourism products and red tourism routes designed around revolutionary history. Strive to achieve the goal of improving the overall grade of red tourism products and building a famous red tourism brand with market influence.
(3) Seize the opportunity of the times and policies to publicize red tourism and inherit red cultural genes. The province continues to build and strengthen the red tourism brand of "Million Miles of Red Road". Relevant departments actively carry out research on red tourism and study the census report in this area in the

province. In the process of designing the red tourism boutique itinerary with the theme of "Retracing the Long March Road", creative points and innovation points have been continuously added. At the same time, red tourism can be developed in combination with policies such as poverty alleviation tourism and green development. Give play to the leading role of red tourism in poverty alleviation, and incline preferential policies to old revolutionary areas with rich red tourism resources but underdeveloped economy. In addition, realize cross-border integration and innovation, and cultivate people and tourism professionals who can tell red stories. Strengthen the publicity of red tourism through advanced means to make more full use of red tourism resources. All government departments should go hand in hand with people from all walks of life, work together, work together, gather strengths from all parties, integrate various resources, and strive for excellence in the development of red tourism.

4.3 Deficiencies and Prospects

This research mainly collected through CNKI literature, Guangdong red cultural map and official websites of various platforms, and sorted out 480 red tourism resources. Because the collection method is too simple, mainly through the network, lack of on-the-spot inspection and book review, and the incompleteness of the online tourism resource data itself, the obtained data has certain limitations and one-sidedness. Insufficient data can lead to structural biases, which can have an impact on the precision of findings. In the future research, it is necessary to improve the deficiencies in this aspect, and to diversify the collection methods to obtain richer and more detailed information and data.

5 Conflicts of Interest

The authors declare no conflict of interest.

Funding This research was funded by Guangzhou Panyu Polytechnic, grant number 2022SK03.

References

1. Xionggen, Chen: The development of red tourism resources and regional economic development in Changzhutan area of Hunan. Econ. Geogr. **06**, 1042–1045 (2007)
2. Liangquan, Zhang, Wenyue, Tang, Wenming, Li.: Recreational value evaluation of red tourism resources from the perspective of place attachment: taking Shaoshan scenic spot as an example. Econ. Geogr. **42**(04), 230–239 (2022)
3. Jie, Xiong, Jinhe, Zhang, Yan, Zhou, et al.: The spatial and temporal distribution characteristics of red tourist attractions in China. Reg. Res. Dev. **37**(02), 83–88 (2018)

4. Guolei, Chen, Chunyan, Zhang, Jing, Luo, et al.: Spatial distribution pattern of classic red tourism scenic spots in China. Resour. Environ. Arid. Reg.Ns **32**(09), 196–202 (2018)
5. Jianqin, Zou, Qingzhong, Ming, Anle, Liu, et al.: The spatial distribution pattern of classic red tourism attractions in China and the heterogeneity of its influencing factors. J. Nat. Resour. **36**(11), 2748–2762 (2021)
6. Zhao, Wang, Wenjie, Huang: Research on the evolution and formation mechanism of red tourist attractions on the Long March. Econ. Geogr. **41**(11), 209–217 (2021)
7. Yun, Xin, Lucang, Wang, Haiyang, Liu: Centennial of the founding of the party: the history and spatial pattern of the formation of national patriotic education bases. Resour. Environ. Arid. Areas **35**(12), 177–187 (2021)
8. Jianxiong, Tang, Youbei, Li., Lin, Xiao: Spatial pattern and influencing factors of red tourism resources in Hunan Province. J. Hunan Univ. Financ. Econ. **35**(02), 74–82 (2019)
9. Hong, Mu.: Research on the development and utilization of red tourism resources based on SWOT model—taking the Hejiang Shidingshan revolutionary site as an example. J. Hunan Inst. Technol. **21**(03), 35–38 (2021)
10. Jiaying, Lu.: Development and research of red tourism resources in Xuefeng Mountain from the perspective of SWOT. Tourism Overview (Second Half Month) **02**, 124–125 (2020)
11. Quanliang, Xiao: Research on the development and utilization of red cultural resources in Qinghai Province based on SWOT analysis. J. Qinghai Norm. Univ. (Philos. Soc. Sci. Ed.) **41**(04), 81–85 (2019)
12. Junbin, Gao, Jingyi, Li.: The development and research of red tourism resources in Lvliang from the perspective of SWOT analysis. Tourism Survey (Second Half Month) **04**, 135–137 (2019)
13. Xiaomei, Li.: SWOT analysis and development strategy of red tourism resources in Hexi Corridor. Tourism Survey (Second Half Month) **16**, 152–153 (2018)
14. Hui, Chen: SWOT analysis of the development of red tourism resources in Dabie Mountains, Anhui Province—from the perspective of Yuexi region. Econ. Res. Guid. **23**, 117–120 (2015)
15. Renli, Xu.: An overview of the development of red tourism abroad and its enlightenment. J. Hubei Univ. Econ. (Hum.Ities Soc. Sci. Ed.) **6**(09), 48–49 (2009)
16. Renli, Xu.: An overview of foreign red tourism and related activities and its enlightenment. Economist **11**, 192–194 (2018)
17. Dongfang, Yang, Mo, Huang: A brief analysis of foreign literature review research on red tourism. Farm Staff **06**, 285 (2018)
18. Davies, C.: Marketing the revolution, the new anti-capitalism and the attack upon corporate brands. Econ. Aff. **23**(3), 61–62 (2003)
19. Ningying, Chen: Analysis of the countermeasures for the development of red culture in Huizhou based on the integration of culture and tourism. J. Zunyi Norm. Univ. **23**(06), 64–67 (2021)
20. Lingxin, Huang, Chunfang, Dou: Research on the current situation and development countermeasures of coastal red tourism resources in Leizhou Peninsula. Ind. Technol. Forum **18**(07), 19–21 (2019)
21. Reyes, M., Francisco-Fernández, M., Cao, R.: Nonparametric kernel density estimation for general grouped data. J. Nonparametric Stat. **28**(2), 235–249 (2016)
22. Sridharan, S., Tunstall, H., Lawder, R., et al.: An exploratory spatial data analysis approach to understanding the relationship between deprivation and mortality in Scotland. Soc. Sci. Med. **65**(9), 1942–1952 (2007)
23. Xijia, Huang, Lijuan, Song: Analysis of the components and development factors of red tourism resources. J. Nanchang Univ. (Hum. Soc. Sci. Ed.) **44**(05), 53–59 (2013)
24. Shimin, Fang, Lijuan, Deng: Classification and evaluation of red tourism resources. Tourism Res. **5**(01), 36–40 (2013)
25. Guangrui, Zhang: "Red Tourism" from a global perspective. Tourism Sci. **22**(05), 1–6 (2008)

26. Xiong, Luo, Yimin, Li.: Red cultural resources and the construction of moral education paths for cultivating newcomers in the era: taking Guangdong's red cultural resources as an example. J. High. Educ. **09**, 53–55 (2019)
27. Tao, Xie: The current situation and countermeasures of the development of red tourism resources in Guangdong. Hong Guang Angle **02**, 48–51 (2015)

The manufacturer's authorised representative in the EU is Springer Nature Customer Service Centre GmbH, Europaplatz 3, 69115 Heidelberg, Germany. If you have any concerns regarding our products, please contact ProductSafety@springernature.com

Printed and bound by CPI Group (UK) Ltd, Croydon, CR0 4YY
01/10/2026
02225930-0005